MW01673516

Semiconductor Infrared Devices and Applications

Semiconductor Infrared Devices and Applications

Editor

A. G. Unil Perera

MDPI • Basel • Beijing • Wuhan • Barcelona • Belgrade • Manchester • Tokyo • Cluj • Tianjin

Editor
A. G. Unil Perera
Georgia State University
USA

Editorial Office
MDPI
St. Alban-Anlage 66
4052 Basel, Switzerland

This is a reprint of articles from the Special Issue published online in the open access journal *Micromachines* (ISSN 2072-666X) (available at: https://www.mdpi.com/journal/micromachines/special_issues/Semiconductor_Infrared_Devices).

For citation purposes, cite each article independently as indicated on the article page online and as indicated below:

LastName, A.A.; LastName, B.B.; LastName, C.C. Article Title. *Journal Name* **Year**, *Volume Number*, Page Range.

ISBN 978-3-0365-3353-7 (Hbk)
ISBN 978-3-0365-3354-4 (PDF)

© 2022 by the authors. Articles in this book are Open Access and distributed under the Creative Commons Attribution (CC BY) license, which allows users to download, copy and build upon published articles, as long as the author and publisher are properly credited, which ensures maximum dissemination and a wider impact of our publications.
The book as a whole is distributed by MDPI under the terms and conditions of the Creative Commons license CC BY-NC-ND.

Contents

About the Editor

A. G. Unil Perera received a B.S. degree in physics (with first class honors) from the University of Colombo, Sri Lanka, and M.S. and Ph.D. degrees from the University of Pittsburgh. He is currently a Regents' Professor at the Department of Physics and Astronomy, Georgia State University, Atlanta. He is a Fellow of the IEEE and Life Fellow of both SPIE and APS. His research focus is on developing tunable and bias selectable detectors responding from UV to FIR and IR applications in disease detection. He has 11 patents, 4 edited books, 11 invited book chapters, and over 200 publications. He has received various awards from the Outstanding Junior Faculty Award and Outstanding Faculty Achievement Award to the Alumni Distinguished Professor Award at GSU. He is also a member of the editorial board for the IEEE Journal of Electron Device Society and also an IEEE Photonics Society Distinguished Lecturer for 2020–2022.

micromachines

Editorial

Editorial for the Special Issue on Semiconductor Infrared Devices and Applications

A. G. Unil Perera

Department of Physics & Astronomy, Georgia State University, Atlanta, GA 30303, USA; uperera@phy-astr.gsu.edu

Infrared radiation (IR) was accidentally discovered in 1800 by the astronomer Sir William Herschel. While trying to study the visible light spectrum and energy in each component, he discovered a type of invisible radiation in the spectrum that was lower in energy than red light. The thermometer used in his experiment can be considered the very first infrared detector, which is categorized as a thermal detector. The first intentional infrared detector is the thermopile developed by Macedonio Melloni in 1835 [1]. One of the early semiconductor materials used as an infrared device was lead sulfide (PbS) [2]. After the second World War, the interest in infrared devices dramatically increased as it became clear that infrared could be used to obtain images of objects due to their heat emission. A detailed history of infrared detector development is presented by Anthony Roglaski in a review article [3]. This led to the establishment of dedicated research facilities for developing infrared detectors such as the Royal Radar Establishment in Malvern, now the Qinetic in the UK, and the US Army Night Vision Laboratory, now the Night Vision and Electronic Sensors Directorate (NVESD) in VA. This focus on infrared imaging for defense applications fueled the rapid development of the field especially for the three atmospheric windows of short-wave IR (SWIR 0.7–2 μm), mid-wave IR (MWIR 3–5 μm), and long-wave IR (LWIR 8–14 μm), where the atmosphere is relatively transparent. The room temperature (300 K) black body radiation happens to peak at 10 μm, which is in the LWIR range.

For a long time, the most studied infrared detector material was HgCdTe, which was heavily used in military applications for night vision, remote sensing, and infrared astronomy research. Changing the Cd composition allowed for the detector to cover the full wavelength spectrum, covering all three ranges from SWIR and MWIR to LWIR. However, the operating temperature of 77 K was a concern for cost-conscious applications. More recently, after the development of novel thin film growth techniques such as molecular beam epitaxy (MBE) and metal–organic chemical vapor deposition (MOCVD), other materials were studied as possible infrared detectors. These developments led to the fast-paced development of the quantum well [4,5], the quantum dot [6,7], and Type II superlattice [8,9] detectors covering various material systems and multiple wavelength ranges. Most of these types of detectors were specifically geared towards various wavelength ranges and specific material systems were developed for each type of detector. Other detection principals based on physics were also studied [10,11], covering a wider wavelength range and with the possibility of being used with any material, which will be advantageous for materials that are already developed. More recently, the idea of extending the accepted standard wavelength threshold governed by the equation $\lambda_t = \frac{1.24}{\Delta}$ was demonstrated [12]. Here, λ_t is the threshold wavelength limit in μms and Δ is the energy gap in meV. All the above-mentioned detector types are known as photon detectors, which are generally much faster than the detectors categorized as thermal detectors. However, thermal detectors [13] have a broad spectral range and cost advantage, making them useful for most practical applications for which millisecond response times are acceptable. Bolometers [14], which belong to the thermal detectors, have also undergone further development due to the advent of microbolometers. Recent technological advances allowing for microstructure designs

Citation: Perera, A.G.U. Editorial for the Special Issue on Semiconductor Infrared Devices and Applications. Micromachines 2021, 12, 1069. https://doi.org/10.3390/mi12091069

Received: 23 August 2021
Accepted: 26 August 2021
Published: 2 September 2021

Publisher's Note: MDPI stays neutral with regard to jurisdictional claims in published maps and institutional affiliations.

Copyright: © 2021 by the author. Licensee MDPI, Basel, Switzerland. This article is an open access article distributed under the terms and conditions of the Creative Commons Attribution (CC BY) license (https://creativecommons.org/licenses/by/4.0/).

have improved the response time of novel microbolometers, providing cost-effective and reasonably fast infrared detectors for mass production applications. The strong optical absorption in human tissue can, in general, be a limitation for optical imaging used for medical diagnosis. This absorbed energy leads to the thermal expansion of tissue, which can generate ultrasound energy when detected by a transducer and produce images of optical absorption contrast within tissues, now known as photoacoustic imaging [15]. A wide variety of infrared detectors have provided application opportunities for almost all areas of humanity involving security and defense, biomedical, commercial, industrial, and scientific research. In fact, the very first technique for checking COVID-19 used a near IR thermometer, which can detect body temperature without contact, providing a safe, quick and easy way to measure the body temperature.

This Special Issue has seven papers covering various aspects of photon detection techniques. Three papers (the first, third, and fifth in this Special Issue) focus on Type II superlattice (SL) infrared detectors. In antimonide-based III–V materials grown on the GaSb substrate, the epi layers (grown on the GaSb substrate) can be lattice-matched or strained. For example, one can grow a thick layer of bulk $InAs_{0.91}Sb_{0.09}$ alloy, which is lattice-matched to GaSb with no strain in the layer. Similarly, an $InAs_{0.91}Sb_{0.09}$/GaSb superlattice also has no strain in the constituent layers as they all are lattice-matched to the GaSb substrate. Conversely, one can design a superlattice with two constituent layers which are not lattice-matched but maintain the overall strain in the superlattice layer at zero. These are called 'strained' layer superlattices (SLS) because the individual layers are strained (tensile or compressive). For example, in a Ga-free InAs/$InAs_{(1-x)}Sb_x$ superlattice the InAs layer is tensile-strained, while the $InAs_{(1-x)}Sb_x$ layer is compressively strained. The alloy composition (x) and layer thickness are the two parameters used to balance the strain in the superlattice unit cell. Similar to InAs/InAsSb superlattices, InAs/InGaSb superlattices are also strained. All the Sb-based superlattices are not technically strained (e.g., $InAs_{0.91}Sb_{0.09}$/GaSb); however, all the commonly used superlattices to date are strained. The first paper in this Special Issue, by Raphael Müller et al. from the Fraunhofer Institute, discusses the performance comparison of an InAs/GaSb Type II SL IR detector with the HgCdTe detector in a real-time spectroscopic application [16]. Comparison of roughly a decade of progress (of Type II IR detectors) with more than a half century of progress in HgCdTe IR detectors itself gives an indication of the rapid advances in the III–V-material-based infrared devices. The second paper, by Ru Chen et al. of China University of Petroleum in Beijing, discusses a manganite-based perovskite-type oxide heterojunction showing ultraviolet-to-near-infrared photo response up to room temperature [17]. Perovskite-type oxides are complex metal oxides with important applications as electrical, magnetic, and catalytic materials. The third and fifth papers here are based on InAs/InAsSb SLS structures by Gamini Ariyawansa et al. from the Air Force Research Laboratory [18] in Dayton, Ohio, and David Ting et al. from NASA JPL. Ariyawansa et al. discusses a mid-wavelength infrared detector and a focal plane array for high-temperature operations, utilizing the *nBn* architecture in their SLS. The fifth paper from David Ting et al. from JPL provides a discussion on the emergence of the Type II SLS infrared detectors and discusses the advantages, disadvantages, and recent developments [19]. In the last two decades, IR detectors are being specifically introduced in biomedical imaging. The fourth paper is from Hasan Göktaş et al. from Harran University in Turkey and discusses the limits of thermal sensing for microbolometers, proposing a method to improve the thermal sensitivity [20]. The sixth paper by Rayyan Manwar et al. from the University of Illinois in Chicago discusses a novel imaging technique that combines the benefits of optical resolution and acoustic depth of penetration, denoted as photoacoustic imaging [21]. The last paper of the volume, by Hemendra Ghimire et al. from the Georgia State University discusses the concept of a heterojunction infrared detector that can be used with any semiconductor material [22]. They also describe a possible approach to detect longer threshold wavelengths beyond the corresponding energy thresholds, giving rise to the possibility of higher operating temperatures for longer wavelength detectors. Overall, this volume covers the well-developed fast response HgCdTe photon detectors to the more recently developed Type II SLS detectors, reasonably fast microbolometer thermal detectors, up and coming photoacoustic detection and imaging, and concludes with a detection technique using a novel intriguing idea of going beyond the very well-established energy gap threshold wavelength rule.

Funding: This research received no external funding.

Conflicts of Interest: The author declares no conflict of interest.

References

1. Schettino, E. A new instrument for infrared radiation measurements: The thermopile of Macedonio Melloni. *Ann. Sci.* **1989**, *46*, 511–517. [CrossRef]
2. Putley, E.H.; Arthur, J.B. Lead Sulphide—An Intrinsic Semiconductor. *Proc. Phys. Soc. Sect. B* **1951**, *64*, 616–618. [CrossRef]
3. Rogalski, A. History of infrared detectors. *Opto-Electron. Rev.* **2012**, *20*, 279–308. [CrossRef]
4. Coon, D.D.; Karunasiri, R.P.G. New mode of IR detection using quantum wells. *Appl. Phys. Lett.* **1984**, *45*, 649. [CrossRef]
5. Levine, B.F. Quantum-well infrared photodetectors. *J. Appl. Phys.* **1993**, *74*, R1–R81. [CrossRef]
6. Bhattacharya, P.; Mi, Z. Quantum-Dot Optoelectronic Devices. *Proc. IEEE* **2007**, *95*, 1723–1740. [CrossRef]
7. Razeghi, M.; Esaki, L.; von Klitzing, K. (Eds.) *The Wonder of Nanotechnology: Quantum Optoelectronic Devices and Applications*; SPIE: Bellingham, WA, USA, 2013; pp. 1–893.
8. Smith, D.L.; Mailhiot, C. Proposal for strained type II superlattice infrared detectors. *J. Appl. Phys.* **1987**, *62*, 2545–2548. [CrossRef]
9. Ting, D.Z.-Y.; Soibel, A.; Höglund, L.; Nguyen, J.; Hill, C.J.; Khoshakhlagh, A.; Gunapala, S.D. Type-II Superlattice Infrared Detectors. In *Semiconductors and Semimetals*; Elsevier BV: Amsterdam, The Netherlands, 2011; Volume 84, pp. 1–57.
10. Perera, A. Heterojunction and superlattice detectors for infrared to ultraviolet. *Prog. Quantum Electron.* **2016**, *48*, 1–56. [CrossRef]
11. Szmulowicz, F.; Madarasz, F.L. Blocked impurity band detectors—An analytical model: Figures of merit. *J. Appl. Phys.* **1987**, *62*, 2533–2540. [CrossRef]
12. Lao, Y.F.; Perera, A.U.; Li, L.H.; Khanna, S.P.; Linfield, E.H.; Liu, H.C. Tunable hot-carrier photodetection beyond the bandgap spectral limit. *Nat. Photonics* **2014**, *8*, 412–418. [CrossRef]
13. Talghader, J.J.; Gawarikar, A.S.; Shea, R.P. Spectral selectivity in infrared thermal detection. *Light. Sci. Appl.* **2012**, *1*, e24. [CrossRef]
14. Perera, A.G.U. (Ed.) *Bolometers*; IntechOpen: Rijeka, Croatia, 2012; ISBN 978-953-81-0235-09. [CrossRef]
15. Beard, P. Biomedical photoacoustic imaging. *Interface Focus* **2011**, *1*, 602–631. [CrossRef] [PubMed]
16. Müller, R.; Haertelt, M.; Niemasz, J.; Schwarz, K.; Daumer, V.; Flores, Y.V.; Ostendorf, R.; Rehm, R. Thermoelectrically-Cooled InAs/GaSb Type-II Superlattice Detectors as an Alternative to HgCdTe in a Real-Time Mid-Infrared Backscattering Spectroscopy System. *Micromachines* **2020**, *11*, 1124. [CrossRef] [PubMed]
17. Chen, R.; Lu, Z.; Zhao, K. Manganite Heterojunction Photodetector with Broad Spectral Response Range from 200 nm to 2 µm. *Micromachines* **2020**, *11*, 129. [CrossRef] [PubMed]
18. Ariyawansa, G.; Duran, J.; Reyner, C.; Scheihing, J. InAs/InAsSb Strained-Layer Superlattice Mid-Wavelength Infrared Detector for High-Temperature Operation. *Micromachines* **2019**, *10*, 806. [CrossRef] [PubMed]
19. Ting, D.Z.; Rafol, S.B.; Khoshakhlagh, A.; Soibel, A.; Keo, S.A.; Fisher, A.M.; Pepper, B.J.; Hill, C.J.; Gunapala, S.D. InAs/InAsSb Type-II Strained-Layer Superlattice Infrared Photodetectors. *Micromachines* **2020**, *11*, 958. [CrossRef] [PubMed]
20. Göktaş, H.; Gökhan, F.S. Analysis and Simulation of Forcing the Limits of Thermal Sensing for Microbolometers in CMOS–MEMS Technology. *Micromachines* **2019**, *10*, 733. [CrossRef] [PubMed]
21. Manwar, R.; Kratkiewicz, K.; Avanaki, K. Overview of Ultrasound Detection Technologies for Photoacoustic Imaging. *Micromachines* **2020**, *11*, 692. [CrossRef] [PubMed]
22. Ghimire, H.; Jayaweera, P.V.V.; Somvanshi, D.; Lao, Y.; Perera, A.G.U. Recent Progress on Extended Wavelength and Split-Off Band Heterostructure Infrared Detectors. *Micromachines* **2020**, *11*, 547. [CrossRef] [PubMed]

micromachines

MDPI

Article

Thermoelectrically-Cooled InAs/GaSb Type-II Superlattice Detectors as an Alternative to HgCdTe in a Real-Time Mid-Infrared Backscattering Spectroscopy System

Raphael Müller *, Marko Haertelt, Jasmin Niemasz, Klaus Schwarz, Volker Daumer, Yuri V. Flores, Ralf Ostendorf and Robert Rehm

Fraunhofer Institute for Applied Solid State Physics IAF, Tullastraße 72, 79108 Freiburg, Germany; marko.haertelt@iaf.fraunhofer.de (M.H.); jasmin.niemasz@iaf.fraunhofer.de (J.N.); klaus.schwarz@iaf.fraunhofer.de (K.S.); volker.daumer@iaf.fraunhofer.de (V.D.); yuri.flores@iaf.fraunhofer.de (Y.V.F.); ralf.ostendorf@iaf.frauhofer.de (R.O.); robert.rehm@iaf.fraunhofer.de (R.R.)

* Correspondence: raphael.mueller@iaf.fraunhofer.de

Received: 26 October 2020; Accepted: 14 December 2020; Published: 18 December 2020

Abstract: We report on the development of thermoelectrically cooled (TE-cooled) InAs/GaSb type-II superlattice (T2SL) single element infrared (IR) photodetectors and exemplify their applicability for real-time IR spectroscopy in the mid-infrared in a possible application. As the European Union's Restriction of Hazardous Substances (RoHS) threatens the usage of the state-of-the-art detector material mercury cadmium telluride (MCT), RoHS-compatible alternatives to MCT have to be established for IR detection. We use bandgap engineered InAs/GaSb T2SLs to tailor the temperature-dependent bandgap energy for detection throughout the required spectral range. Molecular beam epitaxy of superlattice samples is performed on GaAs substrates with a metamorphic GaAsSb buffer layer. Photolithographic processing yields laterally-operated T2SL photodetectors. Integrated in a TE-cooled IR detector module, such T2SL photodetectors can be an alternative to MCT photodetectors for spectroscopy applications. Here, we exemplify this by exchanging a commercially available MCT-based IR detector module with our T2SL-based IR detector module in a real-time mid-infrared backscattering spectroscopy system for substance identification. The key detector requirements imposed by the spectroscopy system are a MHz-bandwidth, a broad spectral response, and a high signal-to-noise ratio, all of which are covered by the reported T2SL-based IR detector module. Hence, in this paper, we demonstrate the versatility of TE-cooled InAs/GaSb T2SL photodetectors and their applicability in an IR spectroscopy system.

Keywords: InAs/GaSb; T2SL; IR; photodetector; TE-cooled; spectroscopy; RoHS; MCT

1. Introduction

In numerous applications in science and industry, detection of infrared (IR) radiation is indispensable. A wide area of application is IR spectroscopy in the mid-infrared (MIR, 3–12 μm). Since several substances in gaseous, liquid, and solid state of aggregation have their characteristic transitions here, this region, which is sometimes referred to as the "fingerprint region", is therefore clearly relevant for industrial or medical spectroscopy applications and when chemical identification or verification is required [1,2]. For industrial applications, common requirements of the IR detector arise. These can be summarized as: fast response, broadband spectral coverage, linearity, and high signal-to-noise ratio. These requirements can be met by specially designed IR photodetectors.

In an IR photodetector, a signal is generated after photon absorption across the fundamental bandgap of the underlying semiconductor material. The bandgap energy E_g of this material defines the cutoff wavelength of the detector, implying that radiation of longer wavelength cannot be detected. As the performance of IR photodetectors decreases for longer cutoff wavelength, choosing the detector cutoff wavelength based on the requirements of the application is essential. In general, cooling the detector material improves the performance of IR photodetectors. Utmost performance is achievable with expensive cooling with cryogenic liquids or Stirling coolers. However, for most applications, low-cost, small, lightweight IR detector modules are required. In these modules, the detector element is thermoelectrically cooled (TE-cooled) with multistage Peltier elements to a so-called high operating temperature (HOT) in the range between 180 K and 300 K.

So far, the commercialized state-of-the-art material of choice for HOT IR photodetectors is mercury cadmium telluride (HgCdTe or MCT). This is due to MCT featuring both a bandgap energy that is widely tunable in the IR, as well as a top-notch electrooptical performance. By adjusting the cadmium content, MCT allows for the fabrication of IR photodetectors with a cutoff wavelength in and beyond the fingerprint region. Numerous studies dedicated to the development and optimization of HOT MCT IR detectors have been conducted [3,4]. However, the Restriction of Hazardous Substances (RoHS) of the European Union regulates the allowed concentration of mercury and cadmium in electronic devices [5]. It is only due to temporary exemptions that this regulation does not prohibit the use of MCT detectors. Hence, for future applications, alternative, RoHS-compatible detector materials need to be established.

Since III-V semiconductors do not contain RoHS-restricted substances, RoHS-compatible photodetectors can be fabricated from them. For detection in the MIR, bulk III-V semiconductors are only partly suitable. InSb, the binary III-V material with the lowest bandgap energy, can only be utilized for detection up to around 5 μm when cryogenically cooled or up to around 7 micron for uncooled operation, which is insufficient for many applications. The ternary alloy $InAs_{1-x}Sb_x$ allows for bandgap tuning by modification of the composition. This enables bandgap energies that are smaller than the one of InSb. The limits of the bandgap tuning range for $InAs_{1-x}Sb_x$, i.e., the temperature and the composition dependence of the bandgap, were recently re-investigated [6]. As no substrate material exists that allows for lattice-matched growth of $InAs_{1-x}Sb_x$, handling the layer strain is inevitable.

We investigate InAs/GaSb type-II superlattices (T2SLs) that are RoHS-compatible, feature a widely tunable bandgap energy and can be grown lattice-matched to GaSb [7–9]. InAs/GaSb T2SLs consist of alternating layers of InAs and GaSb that are usually grown by molecular beam epitaxy (MBE). Each individual layer is just a few atomic monolayers wide and acts as a quantum well for charge carriers. By quantum mechanical coupling of neighboring quantum well states, electron, and hole minibands are created, respectively. The fundamental bandgap of this artificial bandgap material opens between the lowest electron miniband and the highest hole miniband (see Figure 1a). It can be tuned by altering the width of the InAs and GaSb sublayers. Due to the peculiar type-IIb band alignment between InAs and GaSb, the superlattice bandgap energy can be engineered flexibly for a spectral range roughly corresponding to 3–20 μm, which is equivalent to photon energies from about 60 to 400 meV.

To illustrate the bandgap tuning in InAs/GaSb T2SLs, in Figure 1b a calculation of the bandgap energy in dependence of the superlattice composition based on the superlattice empirical pseudopotential method is shown [10]. Apparently, the InAs sublayer width has the main impact on the bandgap energy. Commonly, the superlattice composition is given in dependence of the sublayer width of InAs and GaSb, which is calculated based on calibrated growth rates and the MBE shutter sequence during T2SL growth. However, in Figure 1b, an As content of 17% is indicated for the GaSb sublayer, which was determined by X-ray diffraction. During the growth of a GaSb sublayer, the chamber atmosphere still contains As due to previously grown InAs sublayers. Since As is the group V component that is preferably incorporated into the layer, this leads to a non-negligible As content in the nominal GaSb sublayer. Details on the MBE growth procedure and the method for the determination of the As content in the GaSb sublayers are given in [10].

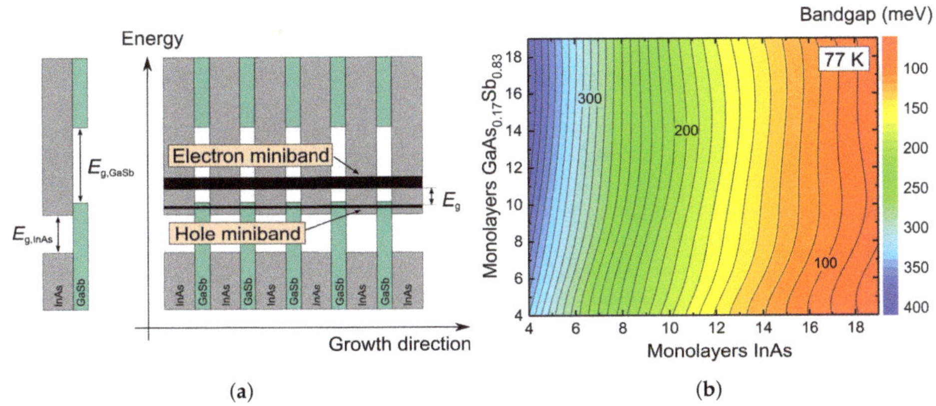

Figure 1. (**a**) Schematic of the type-IIb band alignment between InAs and GaSb and an InAs/GaSb type-II superlattice with lowest electron miniband and highest hole miniband. (**b**) Bandgap energy in dependence of the InAs/GaSb superlattice composition, calculated by the superlattice empirical pseudopotential method [10].

After the proposal to use InAs/GaSb T2SLs for IR detection [11], fundamental research on this material system [12,13] and development of single element detectors [14–16] and detector arrays [17] has intensified in the last decades. Important developmental steps in the field are reviewed in [18].

Activities in research and development of InAs/GaSb T2SLs have mainly focused on high-performance applications at low operating temperatures that require cryogenic cooling. As a result, for low operating temperatures, InAs/GaSb T2SLs emerge as a viable alternative to MCT for IR detectors and IR cameras. For the HOT range, IR detection with InAs/GaSb T2SLs in the longwave infrared was demonstrated [19–22], but dedicated device development and commercialization were never conducted. Now, mainly due to the RoHS, there is renewed interest in InAs/GaSb T2SL IR photodetectors for HOT applications.

Within the last few years, we have worked on the development of InAs/GaSb T2SL single element detectors for the HOT range and demonstrated that they can be combined with the immersion lens technology of VIGO system [23–25]. In this paper, we briefly describe the layout of the detector as a laterally-operated photoconductor, the superlattice and buffer layer growth as well as the detector processing. Then, after the detector is integrated into an IR detector module with a four-stage TE-cooler, which allows for operation at 200 K, we focus on a possible spectroscopy application in which an MCT-based IR detector module could be replaced by a T2SL-based IR detector module.

In addition to the cutoff wavelength, two more detector figures of merit are crucial for the content of this paper. The first is the specific detectivity D^*, which describes the signal-to-noise ratio:

$$D^*(\lambda, f) = \frac{R(\lambda)}{I_n(f)} \sqrt{A_o \Delta f}. \tag{1}$$

D^* depends on the spectral responsivity $R(\lambda)$, the noise current $I_n(f)$ and the bandwidth Δf. It is normalized to the optical detector area A_o. By using a lens to focus incoming radiation, A_o can be increased significantly. The increase depends on the form of the lens and its refractive index n. A hyperhemispheric lens can increase A_o by a factor of n^4 [26]. For backside-illuminated detectors, the lens can be immersed into the substrate material beneath the detector. The second figure of merit is the detector bandwidth that relates to the detection speed. For the device concept under study, the detector bandwidth is inversely proportional to the carrier recombination time. However, the responsivity is proportional to the carrier recombination time. Therefore, there is a trade-off between photosignal and detection speed in photoconductor optimization.

2. Design, Growth, Processing and Module Integration of an IR Detector

The InAs/GaSb T2SL discussed in this paper was grown by molecular beam epitaxy on a 3 inch, n-type, (100)-oriented, 1100 μm thick GaAs substrate after careful calibration of shutter sequences and growth rates. Figure 2a shows the epitaxial layer structure. It consists of two buffer layers, the superlattice absorber layer and a thin superlattice contact layer. The first buffer layer is a metamorphic GaAsSb buffer, in which Sb gradually replaces As over 2 μm layer width. This results in a strain relaxed GaSb-like growth template for the subsequent layers [23]. The second buffer layer consists of 10 μm GaSb. This layer is followed by the superlattice absorber layer, which comprises 750 non-intentionally doped superlattice periods (residually n-type). Each of these periods features 14 monolayers (ML) InAs and 7 ML GaSb. InSb-like interfacial layers were realized between the individual InAs and GaSb sublayers to minimize the relative lattice mismatch to the underlying substrate. In the end, the heavily n-type doped contact layer was grown on top.

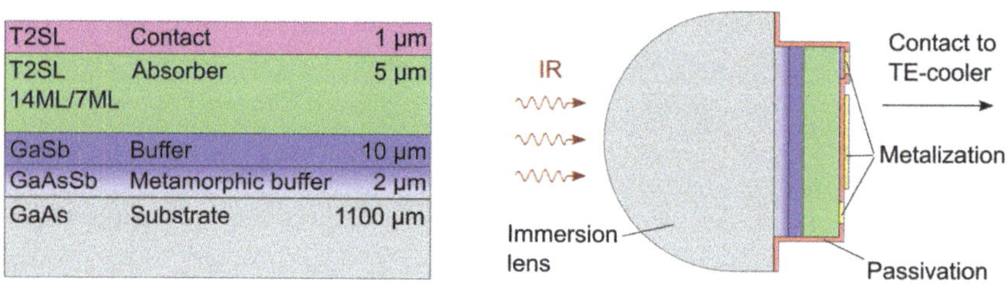

(a) (b)

Figure 2. (**a**) Epitaxial layer structure for the fabrication of laterally-operated InAs/GaSb type-II superlattice (T2SL) detectors on GaAs substrate. (**b**) Schematic of a processed InAs/GaSb T2SL detector that is backside-illuminated through an immersion lens (not to scale).

After growth, standard superlattice layer characterization was performed. A superlattice period length of 7.0 nm was determined by high-resolution X-ray diffraction, which was also used to verify the negligible relative lattice mismatch to the GaSb buffer. Spectral photoluminescence was measured to confirm the intended bandgap energy. At 10 K, a bandgap energy of 143 meV (corresponding to a wavelength of 8.7 μm) was obtained. As the bandgap shrinks for rising temperature, which can be described with the Varshni model [25], the corresponding cutoff wavelength of a detector increases. Hence, this superlattice can absorb radiation throughout a large fraction of the MIR at high operating temperatures.

Photolithographic processing was used to fabricate laterally-operated photoconductors (see Figure 2b). Unlike in most T2SL-based detector concepts, in which the current flows parallel to the superlattice growth direction, in this concept, the current flows mainly perpendicular to the growth direction between two ohmic metal contacts and requires external bias voltage for operation. When radiation of suitable wavelength enters the absorber layer, an additional photoconductivity is generated. The processing steps for detector fabrication included dry etching for structuring of the contact and the absorber layer, dielectric passivation, selective opening of the passivation layer and metalization. In the last step, the mesa front was also metalized. This metalized area acts as a mirror facilitating a double pass of the radiation incident from the backside, which increases the quantum efficiency. The processing sequence has been presented in more detail before [24].

A differing lattice constant of layer and substrate, which is the case for InAs/GaSb T2SLs lattice matched to GaSb on GaAs substrates, may result in an increased density of defects, growth inhomogeneities and a reduced device yield. Our wafer-level device characterization at 200 K suggested that device drop out due to material- or processing-related defects is negligible [24]. As device performance proved to be homogeneous across the wafer, a large device yield would be expected for manufacturing purposes. To allow for immersion of hyperhemispheric microlenses into the substrate, the detectors were processed with a horizontal and vertical pitch of 1480 μm. In this way, more than 1000 detectors could be fabricated per 3 inch wafer—the wafer size used in our study. Assuming the increasingly common 4 inch and 6 inch GaSb substrate diameters, the number of devices per wafer would scale according to the wafer area. For fabrication of detectors without substrate microlenses, the number of detectors per wafer depends on the intended detector size and can be significantly higher.

After processing and the characterization of the T2SLs and the fabricated detectors, the fully processed 3 inch wafers were diced into single element detectors. The module integration of the detector elements was completed in cooperation with VIGO System. In these modules, a T2SL detector 50 μm × 50 μm in size is mounted on top of a four-stage TE-cooler. The detectors feature a hyperhemispheric lens that was immersed into the GaAs substrate. As $n_{GaAs} \approx 3.3$, the lens increases A_o for backside incident radiation by about two orders of magnitude and D^* by one order of magnitude when compared to detector elements without such an immersion lens. Furthermore, the IR detector modules also comprise standard electronics from VIGO System: a fast preamplifier and a TE-cooler controller. These TE-cooled T2SL-based IR detector modules constitute RoHS-compatible turnkey systems.

3. Comparison to MCT

To benchmark the performance of these detectors, we compare the detectivity of MCT-based and T2SL-based photoconductors without immersion lens. They are operated at 210 K with the noise current taken at 20 kHz. In Figure 3, we show the mean value of the detectivity of InAs/GaSb T2SL photoconductors, which we deduced from measurements that were already discussed in [24]. Here, we compare this mean value with specified detectivities of commercial MCT photoconductors from VIGO System for different cutoff wavelengths from 9–13 μm [27]. For detectors with the same cutoff wavelength of 10.6 μm, the detectivity of the MCT photoconductor is less than a factor of two higher than the detectivity of the T2SL photoconductors. Given the brief development of HOT InAs/GaSb T2SL photodetectors in comparison to the longstanding heritage of MCT photodetectors, this is a highly promising result. Doping optimization [25] and increasing the quantum efficiency are expected to further enhance the T2SL detector performance and increase its competitiveness.

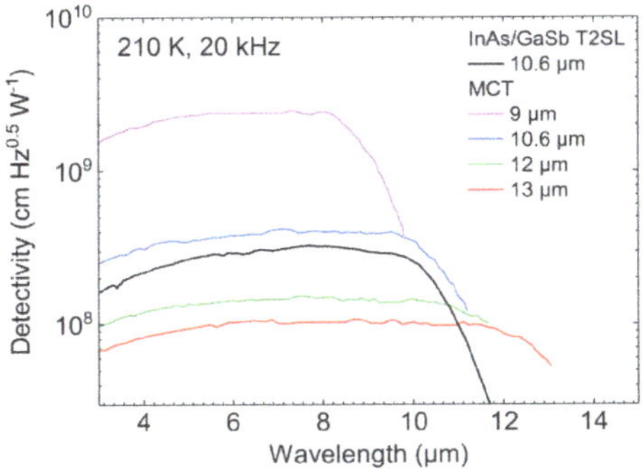

Figure 3. Detectivity of InAs/GaSb T2SL photoconductors (mean value) and commercial mercury cadmium telluride (MCT) photoconductors from VIGO System (guaranteed values) for different cutoff wavelengths at 210 K and 20 kHz [24,27].

As a longer cutoff wavelength implies a lower bandgap and an increased carrier generation, which leads to an increased noise level, the peak detectivity of InAs/GaSb T2SL detectors is expected to drop for longer cutoff wavelengths as it is the case for MCT-based detectors (see Figure 3). As the cutoff of an MCT detector crucially depends on the Cd content in the composition, which becomes more challenging to control precisely and homogeneously towards longer cutoff wavelength, for more elaborate device concepts the device yield drops and in turn the detector price rises. This drawback does not exist for InAs/GaSb T2SLs.

4. Real-Time MIR Backscattering Spectroscopy System

In addition to our development of HOT InAs/GaSb T2SL IR detectors, we realized a demonstrator system for MIR backscattering spectroscopy. The operation principle of the demonstrator exploits the characteristic spectral diffuse reflection of solid chemical substances in the MIR that can be utilized for substance identification. Using a fast spectrally tunable quantum cascade laser (QCL) as the illumination source and a fast photodetector, the system is able to record IR spectra over more than 250 cm^{-1} at rates of 1 kHz and therefore real-time spectroscopy. The high spectral scan speed of the system is ideal for fast changing scenarios or handheld operation as was demonstrated before. Here, we go beyond previous lab demonstrations of the measurement principle [28] as the system can run constantly without user intervention for several hours.

4.1. Setup of the Demonstrator System

The first core component of the system, the IR light source, is an agile wavelength-tunable external cavity quantum cascade laser (EC-QCL) developed by Fraunhofer IAF and Fraunhofer IPMS [28–30]. Its emission wavelength is defined by the deflection of a resonant micro-opto-electro-mechanical system (MOEMS) diffraction grating in Littrow-configuration, which is driven close to the resonance frequency of ~1 kHz (i.e., it harmonically oscillates around its zero-deflection position). Synchronized with the MOEMS oscillation, the EC-QCL is operated in pulsed mode with a pulse length of 100 ns and a repetition rate of about 500 kHz. Due to the resonant nature of the MOEMS scanner, the laser wavelength is continuously tuned and the full spectral range between 1060 cm^{-1} and 1350 cm^{-1} provided by the QCL chip can be scanned in only half a MOEMS period, i.e., ~500 µs. However, typically the IR spectra are constructed from a full MOEMS period, as this increases the spectral resolution. For the parameters mentioned above, one achieves a typical spectral resolution of about 2 cm^{-1} and a spectral broadening per pulse (i.e., per emission wavelength) also of <2 cm^{-1} [27]. These performance parameters allow for spectroscopy on a number of solids and liquids with characteristic bands within the IR fingerprint region. The laser module itself is very compact, as can be seen in Figure 4a. Fraunhofer IAF and IPMS have also developed a non-resonant MOEMS EC-QCL, which allows addressing individual wavelength or (arbitrary) trajectories with scan frequencies of up to several ten hertz in an identical footprint [31].

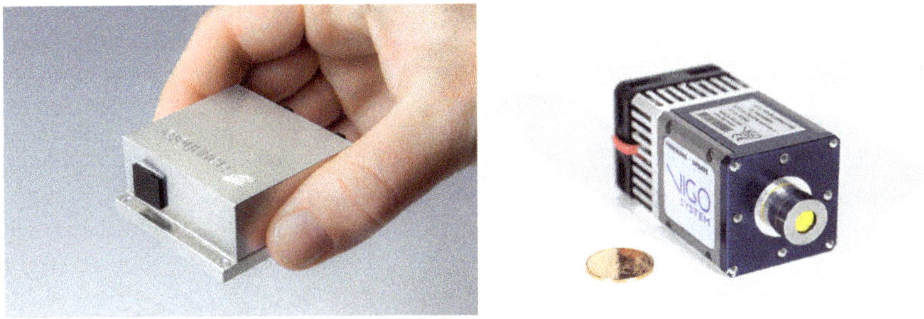

(**a**) (**b**)

Figure 4. Photographs showing (**a**) the compact designs of the external cavity quantum cascade laser (EC-QCL) with micro-opto-electro-mechanical system (MOEMS) diffraction grating and (**b**) the high operating temperature (HOT) T2SL IR detector from Fraunhofer IAF in a detector module from VIGO System.

The second core component of the system, a fast IR photodetector, detects the QCL radiation after it is diffusely reflected by the substance under investigation. The detector was chosen to meet the requirements set by the laser system. These were a MHz-bandwidth, to resolve each individual laser pulse, and a sufficiently long cutoff wavelength, to cover the required spectral range. As in diffuse reflection typical signal intensities are small, a high D^* is also necessary. To achieve a portable and compact system, only TE-cooled detectors were considered. Up until now, only MCT detectors met these requirements and hence an MCT-based IR detector module was initially selected. Its specifications will be presented later, alongside those of the T2SL-based IR detector module. It differs from an MCT-based IR detector module from VIGO only in terms of the employed detector chip. Figure 4b demonstrates the small size of the detector module.

A picture of the demonstrator system and a simplified schematic showing its interior are presented in Figure 5. During the operation of the system, the QCL beam impinges on a continuously rotating sample platform. On this platform, several substances in the form of pills, powders or foils are arranged in small sample compartments. The samples are listed in Table 1.

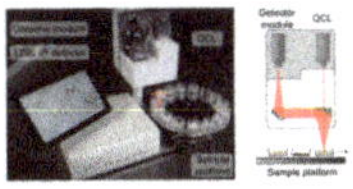

Figure 5. Picture and simplified schematic of the demonstrator system. Backscattering IR spectra are continuously recorded using the tunable EC-QCL and the HOT InAs/GaSb T2SL IR detector. As the sample platform rotates, different substances are illuminated and subsequently identified after comparison with the database.

Table 1. Samples used in the demonstrator system.

Presentation	Sample
pills	Naproxen, Loratadine, Ibuprofen 400 mg, Paracetamol 500 mg, Aspirin 500 mg
powder	Flour, Glucose, Sugar, Lactose, Sweetener, Paracetamol, Aspirin, Caffeine
foil	Kapton, FhG-foil (sticky tape)

As the sample platform rotates, the different substances are sequentially exposed to the incoming QCL beam. The rotational frequency of the sample platform sets the exposure time per substance. In our case, the sample platform rotates with a speed of ~4 rpm, resulting in an exposure time per substance of around ~1 s. After interaction with the respective substance, the laser radiation diffusely backscatters. In the case of the foils, the transmitted light is diffusely backscattered by a plate located below them. Then, the collected portion of the backscattered light is deflected and focused to the fast IR detector. Each single laser pulse is detected, and an IR spectrum is constructed. The substance identification occurs by matching the measured fingerprint spectra to the previously acquired database spectra. The realization of the identification process is described in more detail in the following section.

4.2. System Operation and Database Comparison

Each spectrum measured with the demonstrator system contains a spectral signature, mainly due to the wavelength dependence of the responsivity of the detector. To determine the reflectivity of the different substances under test, the system-dependent spectral signature needs to be corrected for. Hence, at the beginning of the experiment a reference spectrum is acquired with a diffuse scattering plate. It is placed at the same distance as the rotating samples on the sample platform. During operation of the demonstrator system, the measured spectra are always divided by this reference.

With the demonstrator system, IR spectra are continuously recorded at a rate of 1 kHz. Typically, 25 spectra are averaged, corresponding to only 25 ms measurement time. Subsequently, the averaged spectra are compared to a database by using a cross-correlation algorithm that enables substance identification. The database is composed of MIR diffuse reflection spectra (in the case of pills or powders) or transmission spectra (in the case of foils). These spectra were previously acquired with the system itself or a commercial FTIR spectrometer from the same samples. Note that a FTIR measurement takes several minutes in order to achieve a spectral point spacing (~2 cm^{-1}) comparable to our MOEMS EC-QCL-based measurement.

The averaged spectra are continuously compared to the database, while new spectra are still acquired within that time. Hence, no time is lost due to the post-processing of data. Regarding the comparison algorithm, we chose a standard cross-correlation comparison algorithm for simplicity, which is explained in more detail below. The idle time of the system between the recordings of two averaged spectra is sufficient to perform a comparison with the database. The database comprises 15 substances in the given case, which enables a comparison in ~10 ms when using this algorithm. The same algorithm could also be employed for an enlarged database; however, it would be at the cost of a slower database comparison.

In detail, we use the following procedure in our analysis. First, we calculate the normalized cross correlation (NCC) of the averaged spectrum to each database entry. It serves as measure for the similarity between two sets of data. Then, the largest cross-correlation (NCC_{max}) is selected. If NCC_{max} is larger than a threshold value (NCC_{th}), the substance related to the respective database entry is considered as identified. Thereupon, the name of the substance is displayed on the demonstrator screen together with the averaged and the database spectrum. If NCC_{max} is smaller than NCC_{th}, no output is returned. It needs to be mentioned that NCC_{th} is an arbitrary yet fixed number. It is chosen based on the measurement conditions at the beginning of the experiment after an initial test run. It is set as high as possible in order to avoid false positives and as low as possible in order to avoid no returns.

The post-processing of data does not need to interfere with the acquisition of further spectra. To save time, this part can be delegated to different sub-systems or processors on demand, i.e., to a distributed computing architecture. This would also allow a more advanced data processing and analysis. In this context, resolving mixtures into their components or an automatized subtraction of spectral fingerprints from their background are typically of interest. Background subtraction becomes particularly important for the analysis of samples that are not bulk-like, e.g., when a potentially hazardous powder sample on an unknown substrate needs to be identified [32,33].

The presented approach for substance identification or discrimination with the demonstrator setup is solely based on matching IR fingerprint spectra to database spectra that were previously measured for known substances. Therefore, no precise knowledge about the specific nature of the vibrational or vibrational-rotational molecular bands is needed for identification. In fact, the set of substances on the sample platform was chosen arbitrarily. A modified set of substances could also be used as long as the corresponding spectra provide sufficient distinction for discrimination in the MIR.

4.3. Detector Module Interchangeability

To demonstrate the applicability of the T2SL-based IR module for spectroscopy, we replaced the MCT-based IR detector module in the demonstrator system with the T2SL-based IR detector module. The MCT-based module features a two-stage TE-cooled, photovoltaic IR detector that is illuminated via a hemispheric lens resulting in an optical area of 1 mm × 1 mm. This module has been specified with a cutoff wavelength of 10.6 μm, a bandwidth of 100 MHz and a detectivity of 6.8×10^8 cm $\sqrt{\text{Hz}}$/W. The T2SL-based module features a four-stage TE-cooled, photoconductive, 50 μm × 50 μm-sized IR detector that is illuminated via a hyperhemispheric lens, resulting in an optical area of approximately 500 μm × 500 μm. It has been specified with a cutoff wavelength of 9.3 μm, a bandwidth of 10 MHz and a detectivity of 6.7×10^9 cm $\sqrt{\text{Hz}}$/W. The specifications of both detectors are listed in Table 2.

Table 2. Specifications of the two IR detector modules, based on an MCT detector and an InAs/GaSb T2SL detector, respectively.

	MCT	**InAs/GaSb T2SL**
Detectivity	6.8×10^8 cm $\sqrt{\text{Hz}}$/W	6.7×10^9 cm $\sqrt{\text{Hz}}$/W
Bandwidth	100 MHz	10 MHz
Cutoff Wavelength	10.6 μm	9.3 μm
Operating temperature	226 K (2-stage TEC)	200 K (4-stage TEC)
Operation mode	photovoltaic	photoconductive
Optical Area	1 mm × 1 mm	0.5 mm × 0.5 mm
Lens	hemispheric	hyperhemispheric

As both modules feature equal packages and housings from VIGO System, replacing the MCT-based detector module, integrating the T2SL-based IR detector module into the setup and its optical alignment were straightforward. In the following, we report on the operation of the demonstrator system with both IR detector modules.

In Figure 6, exemplary diffuse reflection spectra are shown, which were measured during operation of the demonstrator system on commercial aspirin 500 mg pills and glucose powder with the T2SL-based and MCT-based IR detector module, respectively. The spectra measured with the MCT detector are normalized to their maximum value. The spectra acquired with the T2SL detector are multiplied by a constant, chosen to simplify comparison of the spectra. For both substances, the spectral trends measured with the two IR detectors are well comparable. Clearly, both detectors were able to resolve characteristic spectral features of the aspirin pills and the glucose powder, which allowed for substance identification by database comparison. As all substances on the sample platform (Table 1) have characteristic spectral features in the spectral range coverable with both IR detector modules, the T2SL-based IR detector module was also able to identify them during standard operation after fast comparison with the database in real-time.

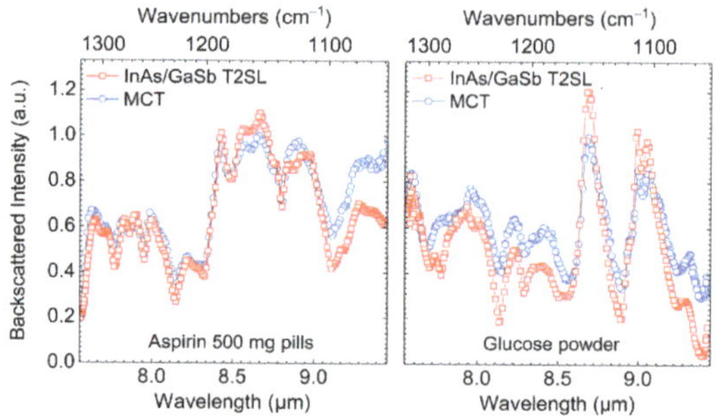

Figure 6. Diffuse IR reflectance spectra obtained from aspirin 500 mg pills and glucose powder with the two IR detector modules featuring a HOT InAs/GaSb T2SL IR detector and a HOT MCT IR detector during operation of the demonstrator system.

4.4. Long-Term Stabilty

We studied the long-term stability of the demonstrator system with the T2SL detector. Following the identification procedure described before, we recorded the identified substance and the calculated NCC as a function of time. More than 50,000 measurements were performed in over 16 h of measurement time. Since the rotation speed of the sample platform is not constant, but rather fluctuates constantly in an uncontrolled manner, on average a new substance was identified every 1.1 ± 0.25 s. As there is no synchronization between the platform and our laser system, for each averaged spectrum, a different area was illuminated and used for analysis.

The results of the long-term stability test of the substance identification are presented in Figure 7. The time evolution is encoded in the figure through the color and size of the dots that are used to represent a single result. Substances for which the reference spectrum was obtained by the FTIR spectrometer are labelled accordingly. Overall, no significant drifts can be observed in the data. However, the distribution of the NCC strongly depends on the substance and varies depending on its form, i.e., foil, pill, or powder. In general, the transmission spectra through the foils show narrower distributions, whereas for the pills the distributions are typically wider. The assumption of our simple model—that each substance can be matched to the database using a single database spectrum—does not necessarily hold for the pills. This relates to the difficulties in solid dose manufacturing to achieve good homogeneity in the blending process. This also broadens the distributions of NCC values in our analysis.

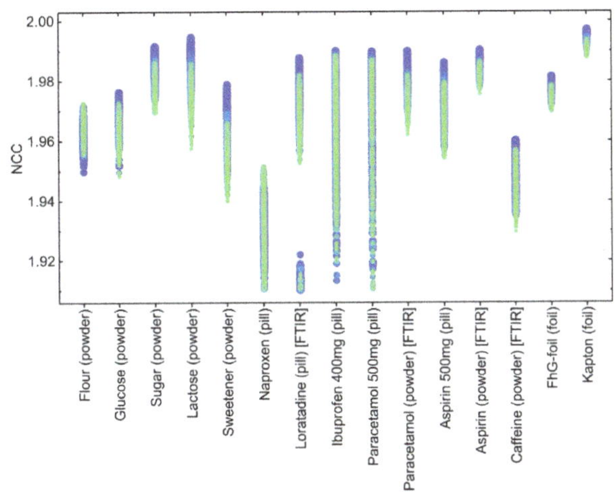

Figure 7. Investigation of the long-term stability of the demonstrator system using the T2SL detector. The time information is encoded in color and size of the dots, i.e., early results are represented by big blue dots, whereas later results are given by smaller and greener dots. Please note the two groups for loratadine, where the lower ones correspond to false assignments.

The detailed analysis of the results of the long-term stability test with more than 50,000 measurements showed that in total 16 samples could not be identified and 68 measurements have been assigned to the wrong substance. Furthermore, 67 of these events correspond to a false assignment to loratadine. Since the corresponding NCCs of these events are smaller and form a separate group in Figure 7, these events could easily be rejected, if a more complex model were used. The 16 missing hits are attributed to ibuprofen 400 mg and the naproxen-based pill, certainly due to a too high NCC_{th} value, which was chosen to be 1.91 in this experiment. In total, the error rate for missing hits is as low as 0.3‰ with the potential to be improved.

5. Discussion

The detector development and the presented application show the potential of TE-cooled T2SL-based IR detector modules for substance discrimination and in a broader scope for IR spectroscopy in general. This potential results from several key properties that the IR detector module exhibits. The first key property is a sufficiently long cutoff wavelength, which is tunable for InAs/GaSb T2SL detectors as it is for MCT detectors. The other key properties are a high detectivity and a bandwidth in the MHz range. A meaningful one-to-one comparison of the two IR detector modules used in the demonstrator system is problematic as they differ in several specifications such as size, operation mode, operating temperature, and cutoff wavelength (see Figure 3). The properties of the T2SL-based IR detector module can be slightly altered by changing the cooling power and hence the operating temperature. Rising the operating temperature of the InAs/GaSb T2SL photoconductor increases the cutoff wavelength and the detector bandwidth but reduces the detectivity. Due to the versatility of the InAs/GaSb T2SL material system and mature device processing at hand, detectors operating in more elaborate device concepts and with properties tailored to a particular application could be realized for the HOT range.

6. Summary

We demonstrated a RoHS-compatible, TE-cooled IR detector module based on an InAs/GaSb T2SL single element detector. For the fabrication of this module, we combined Fraunhofer IAF's expertise in the growth and processing of InAs/GaSb T2SLs with VIGO System's expertise in the fabrication of TE-cooled IR detector modules. This paper shows that this T2SL-based IR detector module and a commercial MCT-based IR detector module can be employed interchangeably in a compact and real-time MIR backscattering spectroscopy system. This system provides a very low error rate of only 0.3‰ in substance differentiation, which can be further improved. Furthermore, we showed that for equal operation mode, operating temperature, cutoff wavelength, and noise frequency, the detectivity of photoconductors based on InAs/GaSb T2SLs and MCT is comparable. This renders InAs/GaSb T2SLs promising for fully RoHS-compatible HOT IR photodetectors.

Author Contributions: Conceptualization, growth, and processing of the IR detector, R.M., J.N., V.D. and R.R.; conceptualization of the laser demonstrator, K.S. and M.H.; methodology, R.M. and M.H.; validation, Y.V.F. and K.S.; data curation, R.M.; writing—original draft preparation, R.M.; writing—review and editing, R.R., Y.V.F. and M.H.; visualization, R.M.; supervision, R.R. and R.O.; project administration, R.R. and R.O.; funding acquisition, R.R. and R.O. All authors have read and agreed to the published version of the manuscript.

Funding: This research was funded by the Horizon 2020 Research and Innovation program under grant agreement no. 688265.

Acknowledgments: The authors thank Peter Holl and Stefan Hugger for developmental work (both with Fraunhofer IAF). We acknowledge the support from VIGO System during the developmental phase of HOT InAs/GaSb T2SL IR photoconductors, including detector characterization (Figure 3). Furthermore, we are grateful to VIGO System for the module integration of our HOT InAs/GaSb T2SL IR detectors.

Conflicts of Interest: The authors declare no conflict of interest. The funders had no role in the design of the study; in the collection, analyses, or interpretation of data; in the writing of the manuscript, or in the decision to publish the results.

References

1. Lambrecht, A.; Schmitt, K. Mid-infrared gas-sensing systems and applications. In *Mid-Infrared Optoelectronics: Materials, Devices, and Applications*; Tournié, E., Cerutti, L., Eds.; Elsevier Science & Technology: Amsterdam, The Netherlands, 2007; pp. 661–715. [CrossRef]
2. Schwaighofer, A.; Brandstetter, M.; Lendl, B. Quantum cascade lasers (QCLs) in biomedical spectroscopy. *Chem. Soc. Rev.* **2017**, *46*, 5903–5924. [CrossRef] [PubMed]
3. Piotrowski, J.; Galus, W.; Grudzien, M. Near room-temperature IR photo-detectors. *Infrared Phys.* **1991**, *31*, 1–48. [CrossRef]
4. Kalinowski, P.; Mikołajczyk, J.; Piotrowski, A.; Piotrowski, J. Recent advances in manufacturing of miniaturized uncooled IR detection modules. *Semicond. Sci. Technol.* **2019**, *34*, 033002. [CrossRef]
5. Directive 2011/65/EU of the European Parliament and of the Council "RoHS-Restriction". Available online: https://eur-lex.europa.eu/legal-content/En/TXT/?uri=celex%3A32011L0065 (accessed on 15 December 2020).
6. Murawski, K.; Gomolka, E.; Kopytko, M.; Grodecki, K.; Michalczewski, K.; Kubiszyn, L.; Gawron, W.; Martyniuk, P.; Rogalski, A.; Piotrowski, J. Bandgap energy determination of InAsSb epilayers grown by molecular beam epitaxy on GaAs substrates. *Prog. Nat. Sci. Mater. Int.* **2019**, *29*, 472–476. [CrossRef]
7. Sai-Halasz, G.A.; Tsu, R.; Esaki, L. A new semiconductor superlattice. *Appl. Phys. Lett.* **1977**, *30*, 651–653. [CrossRef]
8. Sai-Halasz, G.A.; Esaki, L.; Harrison, W.A. InAs-GaSb superlattice energy structure and its semiconductor-semimetal transition. *Phys. Rev. B* **1978**, *18*, 2812–2818. [CrossRef]
9. Sai-Halasz, G.A.; Chang, L.; Welter, J.-M.; Chang, C.-A.; Esaki, L. Optical absorption of In1 xGaxAs/GaSb1 yAsy superlattices. *Solid State Commun.* **1978**, *27*, 935–937. [CrossRef]
10. Masur, J.-M.; Rehm, R.; Schmitz, J.; Kirste, L.; Walther, M. Four-component superlattice empirical pseudopotential method for InAs/GaSb superlattices. *Infrared Phys. Technol.* **2013**, *61*, 129–133. [CrossRef]
11. Smith, D.L.; Mailhiot, C. Proposal for strained type II superlattice infrared detectors. *J. Appl. Phys.* **1987**, *62*, 2545–2548. [CrossRef]
12. Herres, N.; Fuchs, F.; Schmitz, J.; Pavlov, K.M.; Wagner, J.; Ralston, J.D.; Koidl, P.; Gadaleta, C.; Scamarcio, G. Effect of interfacial bonding on the structural and vibrational properties of InAs/GaSb superlattices. *Phys. Rev. B* **1996**, *53*, 15688. [CrossRef]

13. Umana-Membreno, G.A.; Klein, B.; Kala, H.; Antoszewski, J.; Gautam, N.; Kutty, M.N.; Plis, E.; Krishna, S.; Faraone, L. Vertical minority carrier electron transport in p-type InAs/GaSb type-II superlattices. *Appl. Phys. Lett.* **2012**, *101*, 253515. [CrossRef]
14. Johnson, J.L.; Samoska, L.A.; Gossard, A.C.; Merz, J.L.; Jack, M.D.; Chapman, G.R.; Baumgartz, B.A.; Kosai, K.; Johnson, S.M. Electrical and optical properties of infrared photodiodes using the InAs/Ga1-xInxSb superlattice in heterojunctions with GaSb. *J. Appl. Phys.* **1996**, *80*, 1116–1127. [CrossRef]
15. Fuchs, F.; Weimer, U.; Pletschen, W.; Schmitz, J.; Ahlswede, E.; Walter, M.; Wagner, J.; Koidl, P. High performance InAs/Ga1-xInxSb superlattice infrared photodiodes. *Appl. Phys. Lett.* **1997**, *71*, 3251–3253. [CrossRef]
16. Ting, D.; Hill, C.J.; Soibel, A.; Keo, S.A.; Mumolo, J.M.; Nguyen, J.; Gunapala, S.D. A high-performance long wavelength superlattice complementary barrier infrared detector. *Appl. Phys. Lett.* **2009**, *95*, 023508. [CrossRef]
17. Walther, M.; Rehm, R.; Fuchs, F.; Schmitz, J.; Fleißner, J.; Cabanski, W.; Eich, D.; Finck, M.; Rode, W.; Wendler, J.; et al. 256 × 256 focal plane array midwavelength infrared camera based on InAs/GaSb short-period superlattices. *J. Electron. Mater.* **2005**, *34*, 722–725. [CrossRef]
18. Ting, D.; Soibel, A.; Höglund, L.; Nguyen, J.; Hill, C.J.; Khoshakhlagh, A.; Gunapala, S.D. Type-II Superlattice Infrared Detectors. In *Semiconductors and Semimetals*; Gunapala, S.D., Rhiger, D.R., Jagadish, C., Eds.; Academic Press: San Diego, CA, USA, 2011; Volume 84, pp. 1–57.
19. Mohseni, H.; Litvinov, V.I.; Razeghi, M. Interface-induced suppression of the Auger recombination in type-II InAs/GaSb superlattices. *Phys. Rev. B* **1998**, *58*, 15378–15380. [CrossRef]
20. Mohseni, H.; Razeghi, M. Long-wavelength type-II photodiodes operating at room temperature. *IEEE Photonics Technol. Lett.* **2001**, *13*, 517–519. [CrossRef]
21. Mohseni, H.; Michel, E.; Sandoen, J.; Razeghi, M.; Mitchel, W.; Brown, G. Growth and characterization of InAs/GaSb photoconductors for long wavelength infrared range. *Appl. Phys. Lett.* **1997**, *71*, 1403–1405. [CrossRef]
22. Mohseni, H.; Wojkowski, J.; Razeghi, M.; Brown, G.; Mitchel, W. Uncooled InAs-GaSb type-II infrared detectors grown on GaAs substrates for the 8-12-μm atmospheric window. *IEEE J. Quantum Electron.* **1999**, *35*, 1041–1044. [CrossRef]
23. Müller, R.; Gramich, V.; Wauro, M.; Niemasz, J.; Kirste, L.; Daumer, V.; Janaszek, A.; Jureńczyk, J.; Rehm, R. High operating temperature InAs/GaSb type-II superlattice detectors on GaAs substrate for the long wavelength infrared. *Infrared Phys. Technol.* **2019**, *96*, 141–144. [CrossRef]
24. Müller, R.; Niemasz, J.; Daumer, V.; Janaszek, A.; Jureńczyk, J.; Rehm, R. Advances on photoconductive InAs/GaSb type-II superlattice long-wavelength infrared detectors for high operating temperature. *Proc. SPIE* **2019**, *10914*, 1091416. [CrossRef]
25. Müller, R.; Niemasz, J.; Daumer, V.; Rehm, R. Design guidelines for high operating temperature InAs/GaSb type-II superlattice photoconductors for the longwave infrared. (to be published).
26. Piotrowski, J.; Rogalski, A. *High-Operating-Temperature Infrared Photodetectors*; SPIE Press: Bellingham, WA, USA, 2007; p. 28.
27. VIGO System Homepage. Available online: https://vigo.com.pl/produkty/pc-3te/ (accessed on 15 May 2020).
28. Butschek, L.; Hugger, S.; Jarvis, J.; Härtelt, M.; Merten, A.; Schwarzenberg, M.; Grahmann, J.; Stothard, D.M.; Warden, M.; Carson, C.; et al. Microoptoelectromechanical systems-based external cavity quantum cascade lasers for real-time spectroscopy. *Opt. Eng.* **2017**, *57*, 011010. [CrossRef]
29. Grahmann, J.; Merten, A.; Ostendorf, R.; Fontenot, M.; Bleh, D.; Schenk, H.; Wagner, H.-J. Tunable External Cavity Quantum Cascade Lasers (EC-QCL): An application field for MOEMS based scanning gratings. *Proc. SPIE* **2014**, *8977*, 897708. [CrossRef]
30. Ostendorf, R.; Butschek, L.; Hugger, S.; Fuchs, F.; Yang, Q.; Jarvis, J.; Schilling, C.; Rattunde, M.; Merten, A.; Grahmann, J.; et al. Recent advances and applications of external cavity-QCLs towards hyperspectral imaging for standoff detection and real-time spectroscopic sensing of chemicals. *Photonics* **2016**, *3*, 28. [CrossRef]
31. Haertelt, M.; Hugger, S.; Butschek, L.; Schilling, C.; Merten, A.; Schwarzenberg, M.; Dreyhaupt, A.; Grahmann, J.; Rattunde, M.; Ostendorf, R. Advances of MOEMS-based external cavity QCLs. *Proc. SPIE* **2019**, *10926*, 1092613. [CrossRef]

32. Jarvis, J.; Fuchs, F.; Hugger, S.; Ostendorf, R.; Butschek, L.; Yang, Q.; Dreyhaupt, A.; Grahmann, J.; Wagner, J. Hyperspectral image analysis for standoff trace detection using IR laser spectroscopy. *Proc. SPIE* **2016**, *9824*, 98240V. [CrossRef]
33. Jarvis, J.; Haertelt, M.; Hugger, S.; Butschek, L.; Fuchs, F.; Ostendorf, R.; Wagner, J.; Beyerer, J. Hyperspectral data acquisition and analysis in imaging and real-time active MIR backscattering spectroscopy. *Adv. Opt. Technol.* **2017**, *6*, 85–93. [CrossRef]

Publisher's Note: MDPI stays neutral with regard to jurisdictional claims in published maps and institutional affiliations.

© 2020 by the authors. Licensee MDPI, Basel, Switzerland. This article is an open access article distributed under the terms and conditions of the Creative Commons Attribution (CC BY) license (http://creativecommons.org/licenses/by/4.0/).

micromachines

MDPI

Article

Manganite Heterojunction Photodetector with Broad Spectral Response Range from 200 nm to 2 μm

Ru Chen [1], Zhiqing Lu [2] and Kun Zhao [1,*]

1 College of New Energy and Materials, China University of Petroleum (Beijing), Beijing 102249, China; chenlingxs@163.com
2 College of Sciences, China University of Petroleum (Beijing), Beijing 102249, China; hawklzqq@126.com
* Correspondence: zhk@cup.edu.cn; Tel.: +86-10-89734836

Received: 19 December 2019; Accepted: 16 January 2020; Published: 23 January 2020

Abstract: In this paper, we investigate the broad spectral photocurrent properties of the $La_{0.67}Ca_{0.33}MnO_3$/Si (LCMO/Si) heterojunction from 200 nm to 2.0 μm, as the temperature increases from 95 to 300 K. We observed the junction's uniform responsivity in the visible range and five absorption peaks at 940 nm, 1180 nm, 1380 nm, 1580 nm, and 1900 nm wavelengths. The temperature showed effective affection to the photocurrents at absorption peaks and the transition point occurred at 216 K, which was also displayed in the temperature dependence of junction resistance. On the basis of the results, we propose a possible model involving the quantum size effect at the junction interface as the mechanism. This understanding of the infrared photodetection properties of oxide heterostructures should open a route for devising future microelectronic devices.

Keywords: manganite; heterostructure; photodetector

1. Introduction

Oxide semiconductor devices based on the perovskite oxide films, whose properties can be controlled by magnetic field, electric field, and light irradiation, have attracted a great deal of interest. Experiments confirmed that manganite-based perovskite-type oxides have excellent ultraviolet (UV) photoresponse characteristics with ultrafast response-time of picosecond and high sensitivity, which makes this class of materials potentially useful for UV sensor applications [1,2]. Furthermore, manganite heterojunctions offer the features of tunability by magnetic and electric fields, high-sensitivity to light illumination and high carrier mobility, suggesting many possible applications and research directions including information storage, optoelectronics information processing, and advanced sample preparation techniques associated with microstructure modulate research [3–7]. In addition, similar to many optical materials with chemical stability, manganite heterojunctions are insensitive to harsh physical environment such as fluctuations of temperature and pressure, suggesting a potential application of manganite heterojunction photodetectors in harsh environments for the need of oil and gas optics [8–12]. Integrating the perovskite-type transition metal oxides with the silicon (Si)-based semiconductor technology would also introduce the possibility for a multifunctional microelectronic device [13–15].

Si photodetectors have already found wide acceptance for visible light applications, while it has small absorption coefficient in near-infrared (NIR) wavelength range because of the cut off wavelength of ~1100 nm. Now most NIR photodetectors were composed of PbS, PbSe, or InGaAs. The toxic precursors, such as Pb, Se, and As, was usually used to synthesize these materials. It is a meaningful thing to find non-toxic and pollution-free material for photodetector working at NIR wavelength range.

The infrared (IR) spectrum has become an important method to study the lattice distortion and been applied to investigate the photoconductive effect in perovskite manganese oxides, where mid-infrared or far-infrared spectra was used to explain the complex physical process in manganites such as the electronic transition, electron-phonon interaction, coupling between lattice, orbital, and spin, etc., [16–22]. In this paper the photocurrent response spectrum between 200 nm and 2 μm of the heterojunction $La_{0.67}Ca_{0.33}MnO_3$/Si (LCMO/Si) is reported. The temperature dependence of the photocurrent response of the sample was investigated to reveal more information related to the photoelectric response, and selective absorption peaks were observed. The mechanism about the results is also discussed in the paper.

2. Materials and Methods

The LCMO/Si heterojunction was fabricated using the facing target sputtering technique. A 100 nm thickness LCMO layer was grown on a 0.5 mm thick n-type Si (001) wafer. The wafer temperature was kept at 680 °C with the oxygen pressure being 60 mTorr during deposition. Immediately after each deposition, the vacuum chamber was back-filled with 1 atm oxygen gas.

The photocurrent of the sample was detected by the spectral response measurement system, as shown in Figure 1. The system was designed to measure the UV and IR spectral responsivity characteristics of samples in low temperature environment. The operation was automatically controlled, and the system maintained good closure during the measurement process. The selected all-reflected-light-route system, UV, visible light or IR, can be switched automatically with maximum light path coupling efficiency. The diameter of the light spot was 3 mm. The light intensity was calibrated using the spectrum of a commercial UV-100L Si photodiode (from OSI Systems Inc., Hawthorne, CA, USA) and the spectral responsivity was measured by a monochromator.

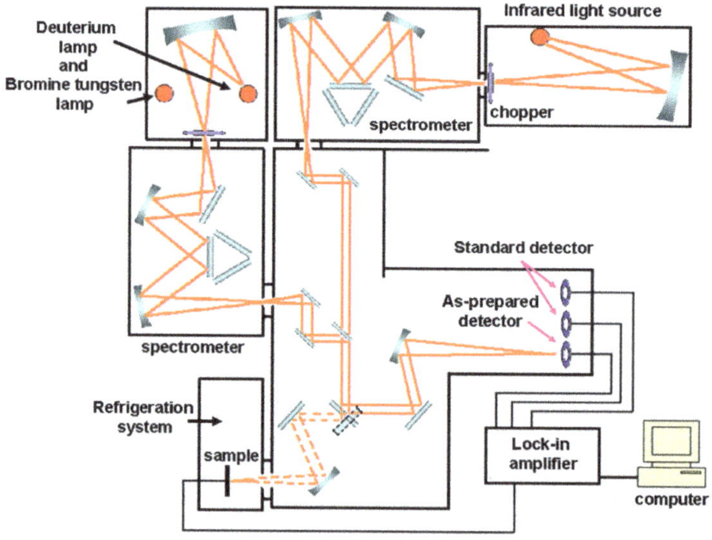

Figure 1. Spectral response measurement system.

The LCMO/Si heterojunction for the photoelectric measurements was cut into 5 × 5 mm and two colloidal silver electrodes were prepared on the LCMO film and Si wafer. The sample was placed in an airtight holder with a quartz window and connected with the spectral response measurement system (Figure 2a). The typical current-voltage curves of the LCMO/Si heterojunction, shown in Figure 2b, were measured in the dark by tuning the applied voltage with a pulse-modulated voltage source at 300 and 60 K. The forward bias was defined as the current flowing from the upper LCMO layer to Si substrate. Thus the diodelike rectification characteristic can be ascribed to the presence of LCMO/Si interfacial potential because of the carrier diffusion.

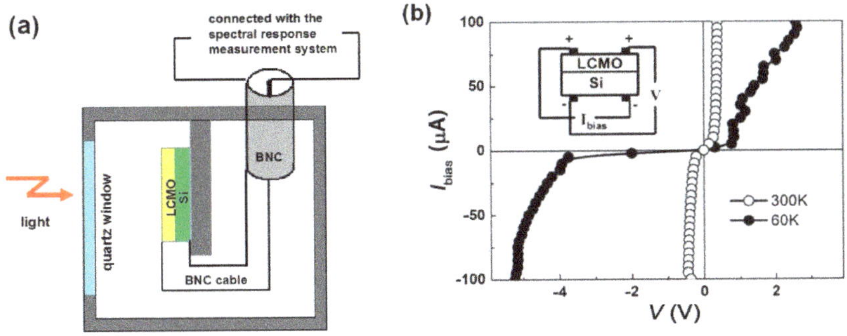

Figure 2. (**a**) The setup of $La_{0.67}Ca_{0.33}MnO_3$/Si (LCMO/Si) heterojunction for the spectral response measurement. (**b**) The current-voltage curves of the LCMO/Si heterojunction at 300 and 60 K.

3. Results and Discussions

The junction resistance R_j in LCMO/Si junction was measured with the temperature. As shown in Figure 3, R_j strongly depends on the bias, e.g., when the bias was turned from 20 μA to −20 μA R_j changed from 166.0 kΩ, 20.0 kΩ, and 15.1 kΩ to 178.1 kΩ, 98.2 kΩ, and 15.6 kΩ at 95 K, 202 K, and 300 K. In addition, taking the bias of 50 μA as an example, R_j decreased slightly from the beginning 95 K to 172 K and had a sharp change from 78 kΩ at 172 K to 8.2 kΩ at 216 K with a corresponding rate of 1.6 kΩ/K. Subsequently R_j maintained small change of about 0.01 kΩ/K till 300 K.

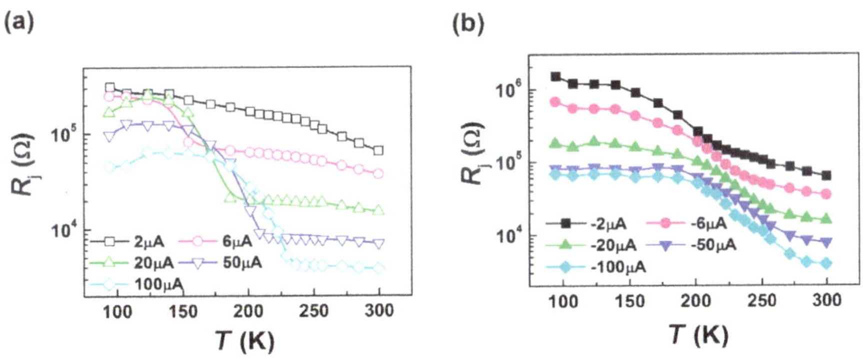

Figure 3. The temperature dependence of junction resistance R_j of a LCMO/Si junction under (**a**) the positive current bias and (**b**) the negative current bias.

Figure 4a displays the photocurrent (*PI*) spectrum of the LCMO/Si junction under zero bias in the wavelength range of 200 nm < λ < 2200 nm. The junction's responsivity was spectrally uniform in the visible range, while five absorption peaks P1, P2, P3, P4, and P5 were observed at λ_1 = 1940 nm, λ_2 = 1180 nm, λ_3 = 1380 nm, λ_4 = 1580 nm, and λ_5 = 1900 nm wavelengths in each temperature because of the absorption characteristics of the LCMO/Si junction, and the peak value decreased with the increase of the wavelength. The temperature dependences of the photocurrent response PI^P at the five absorption wavelengths are shown in Figure 4b. PI^P monotonically increased from 0.0065 A/W, 0.012 A/W, 0.0115 A/W, 0.004 A/W, and 0.002 A/W at 108 K to 0.0122 A/W, 0.017 A/W, 0.0155 A/W, 0.006 A/W, and 0.0025 A/W at a turning point of 216 K and then dropped 0.004 A/W, 0.005 A/W, 0.005 A/W, 0.002 A/W, and 0.001 A/W at 300 K for selected wavelengths of λ_1, λ_2, λ_3, λ_4, and λ_5.

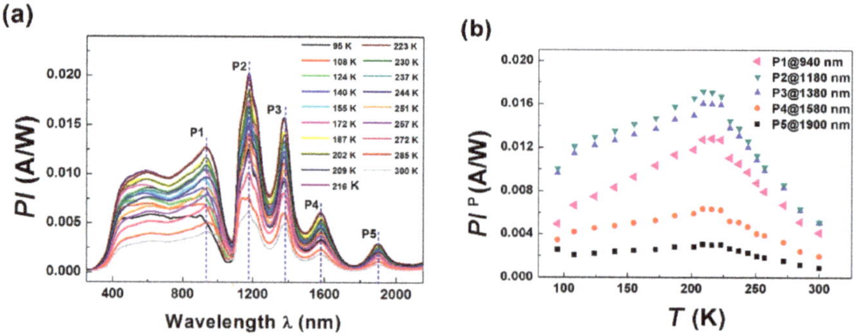

Figure 4. Photocurrent (*PI*) spectrum for various temperatures (**a**) and *PI* peaks (**b**) for 940 nm (P1), 1180 nm (P2), 1380 nm (P3), 1580 nm (P4), and 1900 nm (P5) of a LCMO/Si junction.

Si cannot produce strong absorption features. Most of heterojunctions of manganite-based perovskite-type oxides exhibit the properties of p-n junction and quantum size effect can be produced when the thickness of the potential well material is about 50 nm thick. If quantum size effect occurs, the energy is quantized in the direction of the vertical interface, which will lead to the quantization of energy absorption in the material. Since a thin SiO_2 layer of 3.6 nm thick exists in the LCMO/Si heterojunction [23], a quantum size effect was expected to occur. The interval of adjacent energy levels in an infinite quantum well is described as:

$$\Delta E_{n,n+1} = \pi^2 \eta^2 m_n^{-1} d_w^{-2} (n + 1/2) \propto (n + 1/2) \tag{1}$$

where m_n is the electron effective mass and d_w is the quantum well width. Thus, $(\Delta E_{n,n+1} - \Delta E_{n+1,n+2})$ is independent on n and

$$(\Delta E_{n,n+1} - \Delta E_{n+1,n+2}) - (\Delta E_{n+1,n+2} - \Delta E_{n+2,n+3}) \approx \Delta E_{n,n+1} - 2\Delta E_{n+1,n+2} + \Delta E_{n+2,n+3} \tag{2}$$

As for present five special wavelengths λ_n (n = 1, 2, 3, 4, and 5),

$$(\lambda_1^{-1} - \lambda_2^{-1}) - 2(\lambda_2^{-1} - \lambda_3^{-1}) + (\lambda_3^{-1} - \lambda_4^{-1}) \approx 0.0034 \tag{3}$$

and

$$(\lambda_2^{-1} - \lambda_3^{-1}) - 2(\lambda_3^{-1} - \lambda_4^{-1}) + (\lambda_4^{-1} - \lambda_5^{-1}) \approx 0.0030 \tag{4}$$

Here, the above two similar data suggested that the present model involving the quantum size effect was adopted as the mechanism of IR photocurrent in LCMO/Si.

Noise performance is a critical factor for evaluating a detector. The noise current I_n is about 10^{-4} A/W in dark and is very low compared to the responsivity PI of the LCMO/Si junction when the light was on. The detectivity D^* is determined by the ratio of *PI* and I_n, and $D^* = PI\ (fS)^{1/2}/I_n$, where f is the amplifier frequency bandwidth (500 MHz) and S is the detector area (~7.065 mm^2). Thus D^* is estimated to be about 2.38×10^3 $Hz^{1/2}m$, 2.97×10^3 $Hz^{1/2}m$, 2.97×10^3 $Hz^{1/2}m$, 1.19×10^3 $Hz^{1/2}m$, and 0.59×10^3 $Hz^{1/2}m$ at 300 K for selected wavelengths of λ_1, λ_2, λ_3, λ_4, and λ_5, suggesting that the LCMO/Si junction could be well-suited as an IR detector.

Perovskite-type oxides detectors possess a number of significant characteristics, and are ideally suited to detect small changes in a relatively large background level of incident energy, which can be used over a large spectral bandwidth. Here it has been shown that a specific manganite heterojunction has the ability to be an IR detector since it can produce photocurrent in the IR regime. The devices have a number of important characteristics (low cost, low power, good performance, wide operating range of temperature, a high degree of environmental stability, and reliability) which make them ideal for a range of applications from consumer and commercial to military requirements. LCMO/Si junction is a new material for photodetector fabrication compared to traditional materials. It is anticipated that manganite heterojunction IR detectors will assume an ever growing importance in our society over the next few years.

4. Conclusions

In conclusion, we fabricated a manganite-based heterojunction by depositing a LCMO thin film on the Si substrate. The broad spectral photocurrent effect of the junction was systematically studied in a temperature range from 95 to 300 K. The responsivity of LCMO/Si heterojunction was spectrally uniform in the visible range. Five absorption peaks occurred at 940 nm, 1180 nm, 1380 nm, 1580 nm, and 1900 nm in the IR range, which is explained in terms of a quantum size effect model since an interface existed in the present photodetector. However, relative contributions from individual interface are still not clear and further studies is needed to clarify the *PI* mechanisms.

Author Contributions: Writing—original draft preparation, R.C. and Z.L.; writing—review and editing, K.Z. All authors have read and agreed to the published version of the manuscript.

Funding: This research was funded by the National Nature Science Foundation of China, grant number 11574401.

Acknowledgments: We thank H. K. Wong and Y.C. Kong for the sample preparation in The Chinese University of Hong Kong.

Conflicts of Interest: The authors declare no conflict of interest.

References

1. Jin, K.J.; Lu, H.B.; Zhao, K.; Ge, C.; He, M.; Yang, G.Z. Novel multifunctional properties induced by interface effects in perovskite oxide heterostructures. *Adv. Mater.* **2009**, *21*, 4636–4640. [CrossRef]
2. Lu, Z.Q.; Ni, H.; Zhao, K.; Leng, W.X.; Kong, Y.C.; Wong, H.K. Fast photovoltaic effects tuned by vicinal interface microstructure in manganite-based all-perovskite-oxide heterojunctions. *Appl. Opt.* **2011**, *50*, G23–G26. [CrossRef] [PubMed]
3. Jin, K.J.; Zhao, K.; Lu, H.B.; Liao, L.; Yang, G.Z. Dember effect induced photovoltage in perovskite p-n heterojunctions. *Appl. Phys. Lett.* **2007**, *91*. [CrossRef]
4. Li, X.M.; Zhao, K.; Ni, H.; Zhao, S.Q.; Xiang, W.F.; Lu, Z.Q.; Yue, Z.J.; Wang, F.; Kong, Y.C.; Wong, H.K. Voltage tunable photodetecting properties of $La_{0.4}Ca_{0.6}MnO_3$ films grown on miscut $LaSrAlO_4$ substrates. *Appl. Phys. Lett.* **2010**, *97*. [CrossRef]
5. Ni, H.; Zhao, K.; Xi, J.F.; Feng, X.; Xiang, W.F.; Zhao, S.Q.; Kong, Y.C.; Wong, H.K. Current-pulse-induced enhancement of transient photodetective effect in tilted manganite film. *Opt. Express* **2012**, *20*, 28494–28499. [CrossRef]
6. Ni, H.; Yue, Z.; Zhao, K.; Xiang, W.F.; Zhao, S.Q.; Wang, A.J.; Kong, Y.C.; Wong, H.K. Magnetical and electrical tuning of transient photovoltaic effects in manganite-based heterojunctions. *Opt. Express* **2012**, *20*, A406–A411. [CrossRef]
7. Ni, H.; Zhao, K.; Jin, K.J.; Kong, Y.C.; Wong, H.K.; Xiang, W.F.; Zhao, S.Q.; Zhong, S.X. Nano-domain orientation modulation of photoresponse based on anisotropic transport in manganite films. *Eur. Lett.* **2012**, *97*, 46005. [CrossRef]
8. Zhao, S.S.; Ni, H.; Zhao, K.; Kong, Y.C.; Wong, H.K.; Zhao, S.Q.; Xiang, W.F. Laser induced photovoltaic effects in manganite films for high temperature photodetecting applications in oil and gas optics. *Opt. Commun.* **2013**, *288*, 72–75. [CrossRef]
9. Zhao, S.S.; Ni, H.; Zhao, K.; Xiang, W.F.; Zhao, S.Q.; Kong, Y.C.; Wong, H.K. Manganite heterojunction photodetectors for femtosecond pulse laser measurements. *Opt. Laser Technol.* **2012**, *44*, 1758–1761. [CrossRef]
10. Zhao, S.S.; Ni, H.; Zhao, K.; Zhao, S.Q.; Kong, Y.C.; Wong, H.K. High-sensitivity photovoltaic responses in manganite-based heterojunctions on Si substrates for weak light detection. *Appl. Opt.* **2011**, *50*, 2666–2670. [CrossRef]

11. Ni, H.; Da, S.L.; Zhao, K.; Kong, Y.C.; Wong, H.K.; Zhao, S.Q. Temperature-dependent transport and transient photovoltaic properties of $La_{2/3}Ca_{1/3}MnO_3$/Nb:$SrTiO_3$ heteroepitaxial p-n junction. *J. Appl. Phys.* **2012**, *112*, 023101. [CrossRef]
12. Ni, H.; Da, S.L.; Zhao, K.; Kong, Y.C.; Wong, H.K.; Zhao, S.Q. Transport and transient photovoltaic properties of $La_{0.4}Ca_{0.6}MnO_3$/Nb:$SrTiO_3$ heterojunction at high temperature. *Appl. Phys. A Mater. Sci. Process.* **2012**, *108*, 645–649. [CrossRef]
13. Liu, H.; Zhao, K.; Zhou, N.; Lu, H.B.; He, M.; Huang, Y.H.; Jin, K.J.; Zhou, Y.L.; Yang, G.Z.; Zhao, S.Q.; et al. Photovoltaic effect in micrometer-thick perovskite-type multilayers on Si substrates. *Appl. Phys. Lett.* **2008**, *93*, 171911. [CrossRef]
14. Zhou, N.; Zhao, K.; Liu, H.; Lu, H.B.; He, M.; Zhao, S.Q.; Leng, W.X.; Wang, A.J.; Huang, Y.H.; Jin, K.J.; et al. Enhanced photovoltage in perovskite-type artificial superlattices on Si substrates. *J. Phys. D Appl. Phys.* **2008**, *41*, 155414. [CrossRef]
15. Du, J.; Ni, H.; Zhao, K.; Kong, Y.C.; Wong, H.K.; Zhao, S.Q.; Chen, S.H. Enhanced lateral photovoltaic effect in the p-n heterojunction composed of manganite and silicon by side irradiation for position sensitive detecting. *Opt. Express* **2011**, *19*, 17260–17266. [CrossRef]
16. Okimoto, Y.; Katsufuji, T.; Ishikawa, T.; Urushibara, A.; Arima, T.; Tokura, Y. Anomalous variation of optical spectra with spin polarization in double-exchange ferromagnet: $La_{1-x}Sr_xMnO_3$. *Phys. Rev. Lett.* **1995**, *75*, 109–112. [CrossRef]
17. Kim, K.H.; Gu, J.Y.; Choi, H.S.; Park, G.W.; Noh, T.W. Frequency shifts of the internal phonon modes in $La_{0.7}Ca_{0.3}MnO_3$. *Phys. Rev. Lett.* **1996**, *77*, 1877–1880. [CrossRef]
18. Kaplan, S.G.; Quijada, M.; Drew, H.D.; Tanner, D.B.; Xiong, G.C.; Ramesh, R.; Kwon, C.; Venkatesan, T. Optical evidence for the dynamic Jahn-Teller effect in $Nd_{0.7}Sr_{0.3}MnO_3$. *Phys. Rev. Lett.* **1996**, *77*, 2081–2084. [CrossRef]
19. Calvani, P.; Marzi, G.D.; Dore, P.; Lupi, S.; Maselli, P.; D'Amore, F.; Gagliardi, S.; Cheong, S.W. Infrared absorption from charge density waves in magnetic manganites. *Phys. Rev. Lett.* **1998**, *81*, 4504–4507. [CrossRef]
20. Kim, K.H.; Jung, J.H.; Eom, D.J.; Noh, T.W.; Yu, J.; Choi, E.J. Scaling Behavior of spectral weight changes in perovskite manganites $La_{0.72-y}Pr_yCa_{0.3}MnO_3$. *Phys. Rev. Lett.* **1998**, *81*, 4983–4986. [CrossRef]
21. Sacchetti, A.; Guidi, M.C.; Arcangeletti, E.; Nucara, A.; Calvani, P.; Piccinini, M.; Marcelli, A.; Postorino, P. Far-infrared absorption of La1-xCaxMnO3-y at high pressure. *Phys. Rev. Lett.* **2006**, *96*, 035503. [CrossRef] [PubMed]
22. Pakhira, N.; Krishnamurthy, H.R.; Ramakrishnan, T.V. Optical conductivity of perovskite manganites. *Phys. Rev. B* **2011**, *84*, 085115. [CrossRef]
23. Yue, Z.J.; Zhao, K.; Ni, H.; Zhao, S.Q.; Kong, Y.C.; Wong, H.K.; Wang, A.J. Photo-induced magnetoresistance enhancement in manganite heterojunction at room temperature. *J. Phys. D Appl. Phys.* **2011**, *44*, 095103. [CrossRef]

© 2020 by the authors. Licensee MDPI, Basel, Switzerland. This article is an open access article distributed under the terms and conditions of the Creative Commons Attribution (CC BY) license (http://creativecommons.org/licenses/by/4.0/).

micromachines

Article

InAs/InAsSb Strained-Layer Superlattice Mid-Wavelength Infrared Detector for High-Temperature Operation

Gamini Ariyawansa *, Joshua Duran, Charles Reyner and John Scheihing

Air Force Research Laboratory, Sensors Directorate, Wright-Patterson Air Force Base, OH 45433, USA; joshua.duran.2@us.af.mil (J.D.); charles.reyner.1@us.af.mil (C.R.); john.scheihing@us.af.mil (J.S.)
* Correspondence: Gamini.Ariyawansa.2@us.af.mil

Received: 31 October 2019; Accepted: 19 November 2019; Published: 22 November 2019

Abstract: This paper reports an InAs/InAsSb strained-layer superlattice (SLS) mid-wavelength infrared detector and a focal plane array particularly suited for high-temperature operation. Utilizing the *n*B*n* architecture, the detector structure was grown by molecular beam epitaxy and consists of a 5.5 μm thick *n*-type SLS as the infrared-absorbing element. Through detailed characterization, it was found that the detector exhibits a cut-off wavelength of 5.5 um, a peak external quantum efficiency (without anti-reflection coating) of 56%, and a dark current of 3.4×10^{-4} A/cm^2, which is a factor of 9 times Rule 07, at 160 K temperature. It was also found that the quantum efficiency increases with temperature and reaches ~56% at 140 K, which is probably due to the diffusion length being shorter than the absorber thickness at temperatures below 140 K. A 320 × 256 focal plane array was also fabricated and tested, revealing noise equivalent temperature difference of ~10 mK at 80 K with f/2.3 optics and 3 ms integration time. The overall performance indicates that these SLS detectors have the potential to reach the performance comparable to InSb detectors at temperatures higher than 80 K, enabling high-temperature operation.

Keywords: Infrared detector; strained layer superlattice; InAs/InAsSb; absorption coefficient; barrier detector; high operating temperature

---

1. Introduction

Lower cost, size, weight, and power (C-SWaP) have become a requirement for many infrared imaging systems. A great impact on C-SWaP could be achieved through high operating-temperature (HOT) [1] sensors and focal plane arrays (FPAs), which in turn require developing suitable sensor materials exhibiting high uniformity, high stability, and good electrical and optical properties. One class of materials that has the potential to do just that are III–V, antimony-based, strained-layer superlattices [2] (SLSs), which have already shown impressive results. HOT [1,3] capability has been the primary goal in the mid-wavelength infrared (MWIR) band and a few demonstrations [4] have been already reported. Research groups have explored different detector architectures (*n*B*n* [5], *n*B*p* [6], *p*B*p* [7], XB*n* [8], CBIRDs [9] etc.) and pixel geometries [5,10] across a multitude of SLS designs [11–16] to mitigate generation-recombination (G-R) current and surface-leakage current, while maximizing electrical/optical properties. Among those, detectors with InAs/InAsSb SLSs incorporated in the *n*B*n* architecture have received special attention due to high carrier lifetime [16–18] and reduced complexity of SLS growth [4]. Since the first demonstration of SLSs for infrared (IR) detectors [12], InAs/Ga(In)Sb SLSs continue to improve [19], while InGaAs/InAsSb SLSs [15,20] have also shown promising results. Researchers are currently addressing the poor hole mobility [21] and carrier localization [22,23] effects in *n*-type SLS to improve the diffusion length. They have also explored *p*-type SLS detectors, but surface passivation leading to surface leakage current [24,25] remains an insurmountable problem. In this paper, the focus is on a detector utilizing n-type InAs/InAsSb SLSs based on the nBn architecture, with the emphasis on HOT capability. Detector design, material characterization, and detector performance are discussed in detail, while FPA fabrication and testing are briefly discussed.

2. Strained-Layer Superlattice (SLS) Design and Detector Structure

Group III–V antimony-based SLSs are periodic structures of thin layers of semiconductor materials typically grown on GaSb substrates, which comprise a band-engineered artificial infrared material. In the InAs/InAsSb SLS reported here, the unit cell consists of 16 ML InAs and 6 ML InAsSb layers and the total unit cell thickness was then adjusted to achieve the desired bandgap and spectral response. The band structure of the superlattice was calculated using NRLMultiband software [26] and the band parameters such as the bandgap and carrier effective masses as well as material properties such as absorption coefficient were obtained. Figure 1a illustrates the conduction band (CB) and valence band (VB) profile of the bulk constituents of the superlattice along with superlattice minibands (HH1 and C1) and its bandgap (E_g). The electron and hole probability distributions are also shown, indicating nearly free electrons in the C1 miniband and heavily confined holes in the HH1 miniband. This is a typical feature of InAs/InAsSb SLSs and one can optimize the design [15] to maximize the electron and hole wave function overlap in order to maximize the absorption coefficient and vertical hole mobility. The designed value of E_g is 234 meV (5.3 μm) at 80 K in order to cover the entire 3–5 μm atmospheric band (MWIR).

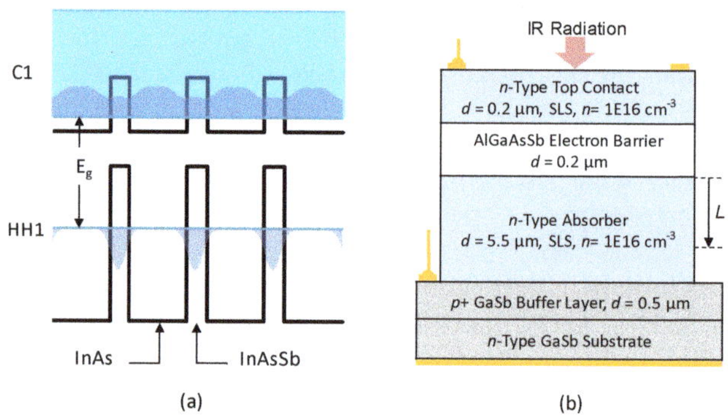

Figure 1. (**a**) InAs/InGaAs strained-layer superlattice (SLS) design showing the conduction and valence band profile for the bulk material as well as the valence and conduction minibands of the superlattice. The superlattice bandgap is also indicated as E_g. (**b**) Structure of the nBn detector consisting of SLS layers as the absorber and contact layers and a bulk AlGaAsSb layer as the electron barrier.

The SLS shown in Figure 1a was incorporated into an nBn architecture in order to build a detector. Figure 1b shows the complete structure of the nBn detector grown on GaSb substrate by molecular beam epitaxy at a commercial foundry. The active elements in the structure include a 5.5 μm thick n-type SLS absorber, a 0.2 μm thick AlGaAsSb electron barrier layer, and a 0.2 μm thick n-type SLS top contact layer. For single element device characterization, mesas were fabricated using standard photolithography, wet chemical etching, and metallization processes. A fully fabricated mesa device is illustrated in Figure 1b. In addition, a metallic mirror was also deposited on the backside of the wafer; this mirror provides a double pass optical geometry under front-side illumination, which approximates the performance of a backside-illuminated FPA.

3. Characteristics of Detectors

Material and device characterization was performed at a range of temperatures from 78 to 300 K. The absorption coefficient (α) was determined from transmission and reflection measurements using a

Fourier transform infrared (FTIR) spectrometer; the details of the method are discussed elsewhere [27]. The absorption coefficient spectra for the superlattice reported here is shown in Figure 2 at 78 and 300 K. Also indicated in the Figure 2 is the cut-off wavelength (~5.2 μm) corresponding to the inflection point of the spectrum (also the edge of the Urbach [28] tail). This value is very close to the designed bandgap of the superlattice (5.3 μm). The value of α at 4 μm and 300 K is 3081 cm^{-1} and the average α in the 3–5 μm band is 3461 cm^{-1}. Furthermore, the cut-off wavelength shifts to about 6.5 μm at 300 K, as expected. These spectra were used for calculating the absorption efficiency in the absorber for comparison against the measured quantum efficiency (QE) of the detector, which will be discussed later.

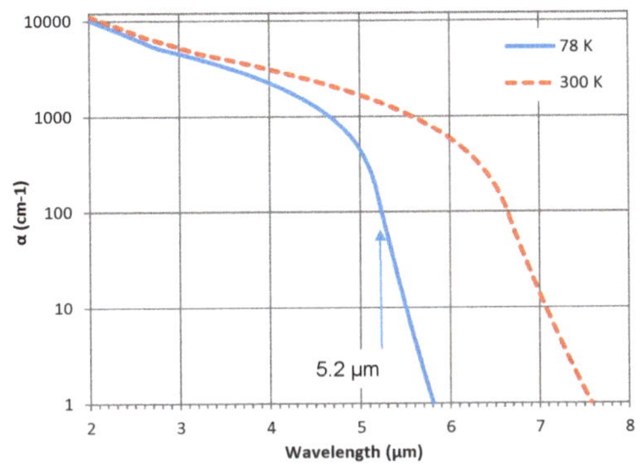

Figure 2. Measured absorption coefficient spectra of the InAs/InAsSb SLS at 78 and 300 K.

The fully processed detectors were packaged in leadless chip carriers and wire-bonded to the chip carrier leads to make electrical contacts. Dark current and photocurrent measurements were carried out after mounting the packaged devices in a liquid nitrogen pour-filled dewar. The dark current-voltage-temperature (IVT) characteristics measured in the 80–240 K temperature range are shown in Figure 3a. Here, the bias polarity is defined as negative (positive) when a negative (positive) voltage is applied on the top contact. The *n*B*n* detector is operated under negative bias where majority electrons flowing from the top contact to the bottom contact are blocked by the electron barrier, while photo-generated minority holes are collected at the top contact. Figure 3b shows the variation of the dark current density (J_d) under –0.2 V bias with temperature (T) with a linear fitting to the experimental data at temperatures higher than 160 K. Based on the slope of J_d/T^3 vs. $1/T$ plot, the activation energy was calculated to be approximately 203 meV. This value is very close to the bandgap of the SLS at ~200 K (confirmed by the spectral cut-off discussed later). It also confirms that the dark current at T > 160 K is diffusion limited. At lower temperatures, the dark current of SLS-based *n*B*n* detectors is typically limited by generation/recombination (G-R) current which is characterized by an activation energy of approximately half the bandgap. This is confirmed by an activation energy of ~115 meV, obtained from the slope of J_d/T^3 vs. $1/T$ plot for T < 100 K, which is approximately half of the bandgap.

Photocurrent was measured using a calibrated blackbody and a set of notch filters at a few specific wavelengths. Then, the quantum efficiency was calculated through radiometric analysis. The resulting quantum efficiency of the detector and its variation with bias at 80 K and 3.4 μm are shown in Figure 4. It should be noted that this is the external quantum efficiency of the detector measured without using an antireflection (AR) coating. From Figure 4, it appears that the detector reaches 90% of max QE (i.e. the turn-on voltage) at a bias of ~ –0.2 V and QE increases slowly when the bias voltage magnitude is increased further. Unlike for a homojunction diode, *n*B*n* detectors are not expected to operate at 0 V, as there is no built-in field in the structure. However, pushing the operating bias voltage near 0 V is preferred, which can be done through optimization of the barrier band alignment and doping. The turn-on voltage at 80 K is reasonably small (200 mV), which decreases with increasing temperature [20].

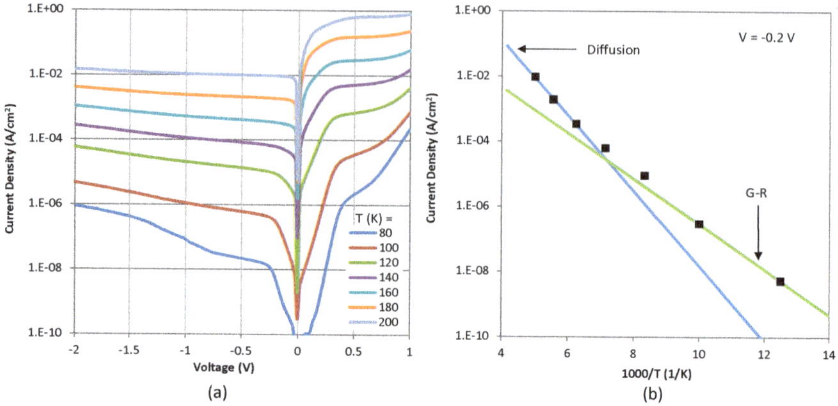

Figure 3. (**a**) Dark current-voltage-temperature (IVT) characteristics of a 400 × 400 μm size *n*B*n* detector and (**b**) Arrhenius plot at a bias voltage of −0.2 V. The linear fit to the current at T > 160 K yields an activation energy of 203 meV.

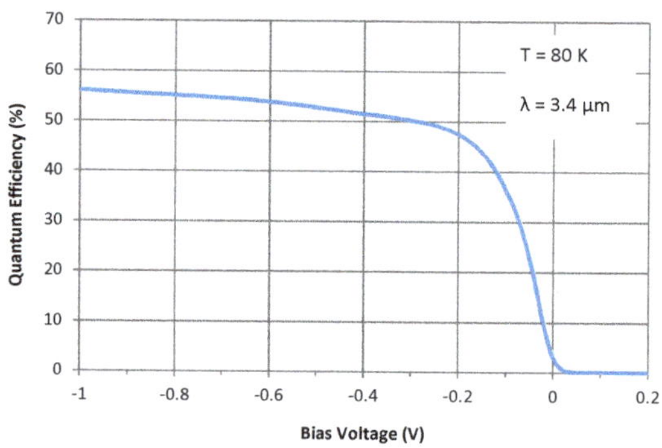

Figure 4. Variation of the external quantum efficiency (no AR coating) of the detector with bias at 80 K and 3.4 μm.

Another important detector characteristic is the spectral response, which is typically measured using a spectrometer. An FTIR spectrometer was used to measure the relative spectral response of this device, which was scaled to spectral QE using the calibrated blackbody measurements. The spectral QE and its variation with temperature under a bias of −0.2 V is shown in Figure 5a. As observed, the cut-off wavelength (inflection point) at 80 K is approximately 5.25 μm (236 meV), which is extremely close to the designed bandgap of the superlattice (234 meV). If the 50% of peak QE is considered, the corresponding wavelength at the band edge is approximately 5.12 μm. This value will be considered as the cut-off for a comparison of the dark current against that of mercury cadmium telluride (MCT) detectors described by Rule 07 [29], as discussed in the next section. As the temperature is increased, the cut-off wavelength increases as expected, but the increase in peak QE is not ideal. As shown in Figure 5b, the magnitude of the peak QE (at 4.2 μm) increases with temperature in the 80–140 K range. At 140 K and beyond, QE saturates, indicating that the QE is likely absorption-limited at these temperatures. This value of QE will be compared with the maximum theoretical value, equal to the absorption efficiency, in the following section. The overall result indicates that this detector's QE peaks at T > 140 K, making it suitable for HOT detectors.

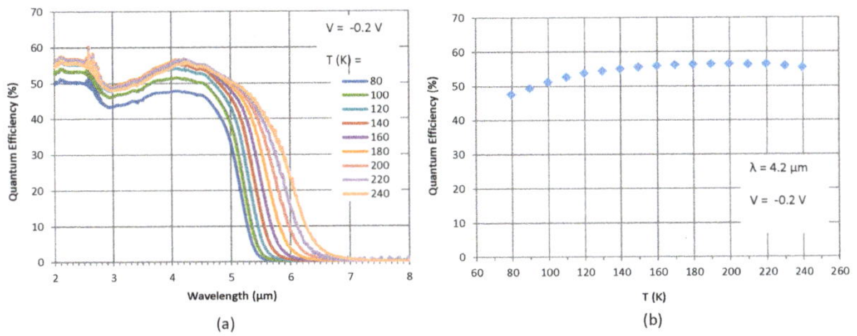

Figure 5. (**a**) Spectral quantum efficiency of the detector at –0.2 V at various temperatures and (**b**) variation of the peak quantum efficiency with temperature at 4.2 µm under –0.2 V.

4. Discussion

Using the experimentally determined absorption coefficient spectra shown in Figure 2, the total absorption in the 5.5 µm thick absorber was calculated [27] using the transfer matrix method under frontside illumination. This calculated absorption efficiency corresponds to the maximum theoretical quantum efficiency of the detector. A comparison of the absorption at 78 and 300 K is compared with the detector QE measured at 80 and 240 K, as shown in Figure 6. It was not possible to measure both absorption efficiency and the quantum efficiency at the same temperature, therefore, close values for the temperature were chosen for this comparison. It is clear that the value of QE at high temperatures is in good agreement with the absorption efficiency. The minor discrepancy observed in the overall spectral shape and the band edge could be due to two reasons: (i) the temperature difference (77 K vs. 80 K and 240 K vs. 300 K), impacting the bandgap and the cut-off wavelength, and (ii) optical resonant effects in the structure, which are very sensitive to the refractive index of the layers in the structure. Currently, the model considers real refractive index values reported in the literature, however, the actual values of the refractive index for the layers in the structure should be measured at corresponding temperatures and used in the model in order to further improve the simulated results.

Figure 6 also confirms that the maximum QE, shown in Figure 5b, is very close to the maximum theoretical QE, i.e. collection efficiency is near unity. However, for T < 140 K, the QE decreases as the temperature is decreased, indicating collection efficiency is less than 1 at these temperatures. Assuming that the diffusion length, L, is lower than the absorber thickness (= 5.5 µm), the absorption in a portion of the absorber with a thickness equal to L measured from the barrier/absorber interface, as indicated in Figure 1b, was calculated. Then, the value of L that gives the best fit between the absorption and QE was determined. At T ~80 K, this value was found to be ~4.8 µm. In other words, QE measured at 80 K corresponds to the collection of carriers generated within a ~4.8 µm region of the absorber. In Figure 6, the absorption efficiency spectra at 78 K correspond to L = 4.8 and 5.5 µm and the spectrum when L = 4.8 µm fits reasonably well with the quantum efficiency spectrum measured at 80 K. Moreover, when the temperature is increased from 80 to 140 K, L increases from 4.8 µm to 5.5 µm, respectively. While this is one straightforward way to explain the QE dependence on temperature, there could be other effects leading to the same observation such as variation in the barrier band alignment to the absorber with temperature and recombination of trapped holes at the interfaces [4].

As of today, MCT technology is still the leading technology for HOT detectors, while SLS technology has become a viable competitor. Therefore, it is worthwhile comparing the dark current between state-of-the-art MCT detectors described by Rule 07 [29] and the SLS detector reported in this paper. Defining the cut-off wavelength as the wavelength near the band edge corresponding to 50% of the peak QE (see Figure 5a), the cut-off wavelength values at different temperatures were determined and the Rule 07 dark current corresponding to those cut-off values and temperatures were calculated. It was then found that the dark current of this SLS detector is approximately a factor of 9, 4, and 3 higher than that of Rule 07 at 160, 180, and 200 K temperatures, respectively.

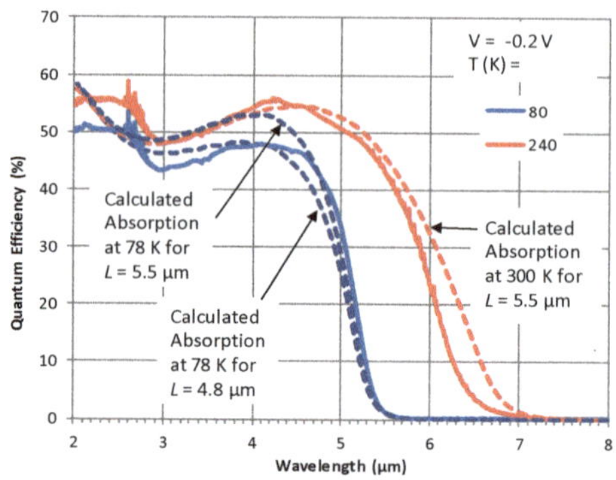

Figure 6. Comparison of the quantum efficiency measured at 80 and 240 K against the absorption efficiency calculated at 78 K (for L = 4.8 and 5.5 μm) and 300 K (for L = 5.5 μm), indicating a good agreement in the quantum efficiency values as well as the spectral shape. The highest temperature for quantum efficiency data available is 240 K and it was chosen to compare against the absorption at 300 K.

The dark current and quantum efficiency of the detector reported in this paper are comparable to similar InAs/InAsSb SLS detectors recently reported in the literature. Ting et al. [4] have reported an InAs/InAsSb SLS nBn detector with a quantum efficiency of ~52% and dark current of 9.6×10^{-5} A/cm^2 (a factor of ~4.5 higher than Rule 07) at ~157 K. With a detailed analysis of dark current characteristics, Rhiger et al. [30] have reported a similar InAs/InAsSb SLS nBn detector exhibiting a dark current 5 times higher than Rule 07. Furthermore, a comprehensive review of antimony-based detectors has been reported by Rogalski et al. [2] With this level of dark current performance and external quantum efficiency >50%, it can be predicted that SLS detectors are well within the reach of performance of InSb detectors but at high temperatures, promising as a candidate for HOT detectors.

To demonstrate the imaging performance, a 320 × 256 detector array with 30 μm pitch was fabricated, flip-chip bonded to a commercial readout integrated circuit chip (FLIR ISC9705), and tested to obtain performance metrics. As shown in Figure 7, the FPA exhibits promising results, including a median noise-equivalent temperature difference (NEDT) of 10 mK. This FPA also showed good uniformity and image quality up to about 140 K. Furthermore, these performance metrics agree with the characteristics measured at the single element detector level, discussed earlier in this paper.

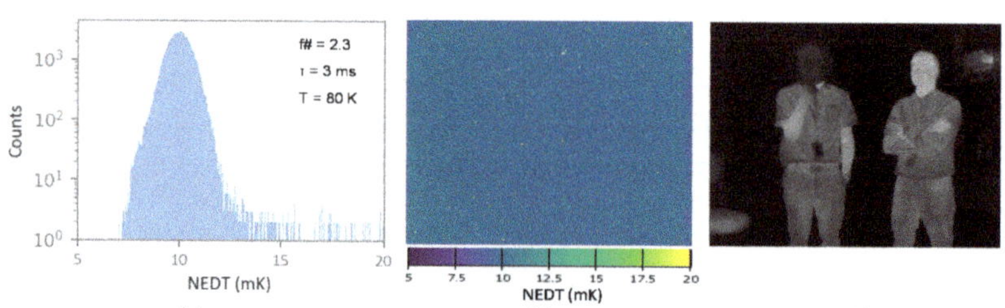

Figure 7. (**a**) Noise-equivalent temperature difference (NEDT) histogram of a 320 × 256 focal plane array (FPA) at 80 K; (**b**) NEDT operability map; and (**c**) an image taken at 80 K with f/2.3 optics and a 3 ms integration time.

5. Conclusions

A MWIR nBn detector designed using InAs/InAsSb SLS was reported. Detector characteristics were measured and analyzed with an emphasis on high temperature operation. At 160 K, this detector exhibits dark current of 9 times Rule 07 and peak quantum efficiency of 56% (~84% of internal quantum efficiency). The turn ON voltage is at or below −200 mV over the full temperature range. It was estimated that the diffusion length of the SLS is approximately 4.8 μm at 80 K, which increases to a value comparable to the absorber thickness (5.5 μm) when the temperature is increased to 140 K. Comparing the calculated absorption efficiency and the measured detector quantum efficiency, it was possible to conclude that the detector exhibits nearly 100% collection at temperatures higher than 140 K. While the performance metrics reported here do not meet those of InSb detectors yet, SLS technology continues to improve with the promise that it has the potential to deliver future HOT detectors required for many applications.

Author Contributions: Conceptualization, G.A., J.D., C.R. and J.S.; methodology, G.A., J.D., C.R. and J.S.; software, G.A. and J.D.; validation, G.A., J.D., C.R. and J.S.; formal analysis, G.A. and J.D.; investigation, G.A. and J.D.; resources, G.A., J.D., C.R. and J.S.; data curation, G.A. and J.D.; writing—original draft preparation, G.A.; writing—review and editing, J.D., C.R. and J.S.; supervision, C.R. and J.S.; project administration, J.S.; funding acquisition, J.S.

Funding: This work was funded by the Air Force Research Laboratory, Sensors Directorate under project "III–V Focal Plane Array Development Using Novel Superlattices".

Acknowledgments: The authors would like to acknowledge the support from the scientists/engineers at the Naval Research Laboratory (NRL) by providing us with the NRL MULTIBANDS software.

Conflicts of Interest: The authors declare no conflict of interest. The funders had no role in the design of the study; in the collection, analyses, or interpretation of data; in the writing of the manuscript; or in the decision to publish the results.

References

1. Martyniuk, P.; Rogalski, A. HOT infrared photodetectors. *Opto Electron. Rev.* **2013**, *21*, 239–257. [CrossRef]
2. Rogalski, A.; Martyniuk, P.; Kopytko, M. *Antimonide-Based Infrared Detectors–A New Perspective*; SPIE Press: Bellingham, WA, USA, 2018.
3. Pour, S.A.; Huang, E.K.; Chen, G.; Haddadi, A.; Nguyen, B.M.; Razeghi, M. High operating temperature midwave infrared photodiodes and focal plane arrays based on type-II InAs/GaSb superlattices. *Appl. Phys. Lett.* **2011**, *98*, 143501. [CrossRef]
4. Ting, D.Z.; Soibel, A.; Khoshakhlagh, A.; Rafol, S.B.; Keo, S.A.; Höglund, L.; Fisher, A.M.; Luong, E.M.; Gunapala, S.D. Mid-wavelength high operating temperature barrier infrared detector and focal plane array. *Appl. Phys. Lett.* **2018**, *113*, 021101. [CrossRef]
5. Maimon, S.; Wicks, G.W. nBn detector, an infrared detector with reduced dark current and higher operating temperature. *Appl. Phys. Lett.* **2006**, *89*, 151109. [CrossRef]
6. Klem, J.F.; Kim, J.K.; Cich, M.J.; Hawkins, S.D.; Fortune, T.R.; Rienstra, J.L. *Comparison of nBn and nBp Mid-Wave Barrier Infrared Photodetectors*; SPIE Press: Bellingham, WA, USA, 2010; Volume 7608.
7. Plis, E.A.; Krishna, S.S.; Gautam, N.; Myers, S.; Krishna, S. Bias Switchable Dual-Band InAs/GaSb Superlattice Detector With pBp Architecture. *Photonics J. IEEE* **2011**, *3*, 234–240. [CrossRef]
8. Klipstein, P. "XBn" barrier photodetectors for high sensitivity and high operating temperature infrared sensors. *SPIE Proc.* **2008**, *6940*, 69402U.
9. Ting, D.Z.Y.; Hill, C.J.; Soibel, A.; Keo, S.A.; Mumolo, J.M.; Nguyen, J.; Gunapala, S.D. A high-performance long wavelength superlattice complementary barrier infrared detector. *Appl. Phys. Lett.* **2009**, *95*, 023508. [CrossRef]
10. Nolde, J.A.; Jackson, E.M.; Bennett, M.F.; Affouda, C.A.; Cleveland, E.R.; Canedy, C.L.; Vurgaftman, I.; Jernigan, G.G.; Meyer, J.R.; Aifer, E.H. Reticulated shallow etch mesa isolation for controlling surface leakage in GaSb-based infrared detectors. *Appl. Phys. Lett.* **2017**, *111*, 051102. [CrossRef]
11. Chang, L.L.; Esaki, L. Electronic properties of InAs-GaSb superlattices. *Surf. Sci.* **1980**, *98*, 70–89. [CrossRef]
12. Mailhiot, C.; Smith, D.L. Long-wavelength infrared detectors based on strained InAs/$Ga_{1-x}In_xSb$ type-II superlattices. *J. Vac. Sci. Technol. A Vac. Surf. Films* **1989**, *7*, 445–449. [CrossRef]
13. Nguyen, B.M.; Hoffman, D.; Delaunay, P.Y.; Razeghi, M. Dark current suppression in type II InAs/GaSb superlattice long wavelength infrared photodiodes with M-structure barrier. *Appl. Phys. Lett.* **2007**, *91*, 163511–163513. [CrossRef]

14. Aifer, E.H.; Tischler, J.G.; Warner, J.H.; Vurgaftman, I.; Bewley, W.W.; Meyer, J.R.; Kim, J.C.; Whitman, L.J.; Canedy, C.L.; Jackson, E.M. W-structured type-II superlattice long-wave infrared photodiodes with high quantum efficiency. *Appl. Phys. Lett.* **2006**, *89*, 053519. [CrossRef]
15. Ariyawansa, G.; Reyner, C.J.; Steenbergen, E.H.; Duran, J.M.; Reding, J.D.; Scheihing, J.E.; Bourassa, H.R.; Liang, B.L.; Huffaker, D.L. InGaAs/InAsSb strained layer superlattices for mid-wave infrared detectors. *Appl. Phys. Lett.* **2016**, *108*, 022106. [CrossRef]
16. Steenbergen, E.H.; Connelly, B.C.; Metcalfe, G.D.; Shen, H.; Wraback, M.; Lubyshev, D.; Qiu, Y.; Fastenau, J.M.; Liu, A.W.K.; Elhamri, S.; et al. Significantly improved minority carrier lifetime observed in a long-wavelength infrared III-V type-II superlattice comprised of InAs/InAsSb. *Appl. Phys. Lett.* **2011**, *99*, 251110. [CrossRef]
17. Höglund, L.; Ting, D.Z.; Soibel, A.; Fisher, A.; Khoshakhlagh, A.; Hill, C.J.; Keo, S.; Gunapala, S.D. Minority carrier lifetime in mid-wavelength infrared InAs/InAsSb superlattices: Photon recycling and the role of radiative and Shockley-Read-Hall recombination mechanisms. *Appl. Phys. Lett.* **2014**, *105*, 193510. [CrossRef]
18. Kadlec, E.A.; Olson, B.V.; Goldflam, M.D.; Kim, J.K.; Klem, J.F.; Hawkins, S.D.; Coon, W.T.; Cavaliere, M.A.; Tauke-Pedretti, A.; Fortune, T.R.; et al. Effects of electron doping level on minority carrier lifetimes in n-type mid-wave infrared InAs/InAs1-xSbx type-II superlattices. *Appl. Phys. Lett.* **2016**, *109*, 261105. [CrossRef]
19. Rogalski, A.; Kopytko, M.; Martyniuk, P. InAs/GaSb type-II superlattice infrared detectors: Three decades of development. In *Infrared Technology and Applications XLIII*; SPIE Press: Bellingham, WA, USA, 2017; p. 1017715.
20. Ariyawansa, G.; Reyner, C.J.; Duran, J.M.; Reding, J.D.; Scheihing, J.E.; Steenbergen, E.H. Unipolar infrared detectors based on InGaAs/InAsSb ternary superlattices. *Appl. Phys. Lett.* **2016**, *109*, 021112. [CrossRef]
21. Olson, B.V.; Klem, J.F.; Kadlec, E.A.; Kim, J.K.; Goldflam, M.D.; Hawkins, S.D.; Tauke-Pedretti, A.; Coon, W.T.; Fortune, T.R.; Shaner, E.A.; et al. Vertical Hole Transport and Carrier Localization in InAs/InAsSb Type-II Superlattice Heterojunction Bipolar Transistors. *Phys. Rev. Appl.* **2017**, *7*, 024016. [CrossRef]
22. Rhiger, D.R.; Smith, E.P. Carrier Transport in the Valence Band of nBn III–V Superlattice Infrared Detectors. *J. Electron. Mater.* **2019**, *48*, 6053–6062. [CrossRef]
23. Steenbergen, E.H.; Massengale, J.A.; Ariyawansa, G.; Zhang, Y.H. Evidence of carrier localization in photoluminescence spectroscopy studies of mid-wavelength infrared InAs/InAs1−xSbx type-II superlattices. *J. Lumin.* **2016**, *178*, 451–456. [CrossRef]
24. Huang, E.K.; Hoffman, D.; Nguyen, B.; Delaunay, P.; Razeghi, M. Surface leakage reduction in narrow band gap jour-II antimonide-based superlattice photodiodes. *Appl. Phys. Lett.* **2009**, *94*, 053506. [CrossRef]
25. Sidor, D.E.; Savich, G.R.; Wicks, G.W. Surface Leakage Mechanisms in III–V Infrared Barrier Detectors. *J. Electron. Mater.* **2016**, *45*, 4663–4667. [CrossRef]
26. Lumb, M.P.; Vurgaftman, I.; Affouda, C.A.; Meyer, J.R.; Aifer, E.H.; Walters, R.J. Quantum wells and superlattices for III-V photovoltaics and photodetectors. *SPIE Proc.* **2012**, *8471*, 84710A.
27. Ariyawansa, G.; Steenbergen, E.; Bissell, L.J.; Duran, J.M.; Scheihing, J.E.; Eismann, M.T. Absorption characteristics of mid-wave infrared type-II superlattices. *SPIE Proc.* **2014**, *9070*, 90701J.
28. Urbach, F. The Long-Wavelength Edge of Photographic Sensitivity and of the Electronic Absorption of Solids. *Phys. Rev.* **1953**, *92*, 1324. [CrossRef]
29. Tennant, W.E.; Lee, D.; Zandian, M.; Piquette, E.; Carmody, M. MBE HgCdTe Technology: A Very General Solution to IR Detection, Described by "Rule 07", a Very Convenient Heuristic. *J. Electron. Mater.* **2008**, *37*, 1406. [CrossRef]
30. Rhiger, D.R.; Smith, E.P.; Kolasa, B.P.; Kim, J.K.; Klem, J.F.; Hawkins, S.D. Analysis of III–V Superlattice nBn Device Characteristics. *J. Electron. Mater.* **2016**, *45*, 4646–4653. [CrossRef]

© 2019 by the authors. Licensee MDPI, Basel, Switzerland. This article is an open access article distributed under the terms and conditions of the Creative Commons Attribution (CC BY) license (http://creativecommons.org/licenses/by/4.0/).

micromachines

MDPI

Article

Analysis and Simulation of Forcing the Limits of Thermal Sensing for Microbolometers in CMOS–MEMS Technology

Hasan Göktaş [1,*] and Fikri Serdar Gökhan [2]

1 Department of Electrical and Electronic Engineering, Harran University, Şanlıurfa 63000, Turkey
2 Department of Electrical and Electronic Engineering, Alanya Alaaddin Keykubat University, Kestel, Alanya, Antalya 07450, Turkey; fsgokhan@gmail.com
* Correspondence: hgoktas.gwu@gmail.com; Tel.: +90-414-318-3000

Received: 25 September 2019; Accepted: 25 October 2019; Published: 29 October 2019

Abstract: Room-temperature highly sensitive microbolometers are becoming very attractive in infrared (IR) sensing with the increase in demand for the internet of things (IOT), night vision, and medical imaging. Different techniques, such as building extremely small-scale devices (nanotubes, etc.) or using 2D materials, showed promising results in terms of high sensitivity with the cost of challenges in fabrication and low-noise readout circuit. Here, we propose a new and simple technique on the application of joule heating on a clamped–clamped beam without adding any complexity. It provides much better uniformity in temperature distribution in comparison to conventional joule heating, and this results in higher thermal stresses on fixed ends. This consequently brings around 60.5× improvement in the overall temperature sensitivity according to both theory and COMSOL (multiphysics solver). The sensitivity increased with the increase in the stiffness constant, and it was calculated as 134 N/m for a device with a 60.5× improvement. A considerable amount of decrease in the operation temperature (36× below 383 K and 47× below 428 K) was achieved via a new technique. That's why the proposed solution can be used either to build highly reliable long-term devices or to increase the thermal sensitivity.

Keywords: microbolometer; infrared sensor; complementary metal-oxide semiconductor (CMOS); high sensitivity; temperature sensor; microresonator; MEMS; clamped–clamped beam; thermal detector

1. Introduction

MEMS/NEMS (Micro/Nano-Electro-Mechanical Systems) resonators got tremendous attention in the last decades, especially with the increase in the demand for the internet of things (IOT), biosensors, gas sensors, and infrared (IR) sensing applications (night vision, gas detection, medical imaging, etc.). Photon detectors [1,2] and microbolometers [3,4] are the two widely used and well-known competitors in building IR sensors. Photon detectors suffer from the requirement of cryogenic cooling, intrinsic noise, relatively high cost, fabrication complexity, and being bulky and expensive. On the other hand, room-temperature microbolometers not only eliminate all these problems, but can also easily be implemented in widely used CMOS (complementary metal-oxide semiconductor) processes [3]. The working principle of the microbolometer is based on the conversion of incident radiation into heat via a plasmonic absorber, and then conversion of this heat into an electrical signal via a temperature sensor. This electrical signal can be either resistance change (non-resonant) or frequency change (resonant type), depending on the device type. Resonant sensors [5–7], in contrast to non-resonant sensors [8], are the most popular because they offer significant advantages [6], such as high-quality factor of 1 million [9], ultra-low-noise measurement, and highly accurate measurement. The working principle of MEMS-type resonant-based thermal sensors is based on the resonance frequency shift, with respect to change in the temperature.

Many different techniques have been applied to increase the sensitivity of resonant-type microbolometers, where the thermal sensitivity strongly depends on the temperature coefficient of frequency (TCF). Extremely small devices [10,11] have higher sensitivity and a higher TCF, but they have challenges, such as difficulties in fabrication, high density integration, and low-noise readout circuit implementation. Using NEMS offered better sensitivity and TCF by achieving better thermal isolation [6,12,13], but sacrifices low stiffness constant. High quality factor [9,14,15] was achieved to detect a much smaller frequency shift and consequently enables higher thermal sensitivity and a higher TCF, but this method either requires vacuum environment or fabrication complexity or fabrication compatibility problem with CMOS process. Using phase change materials [16] in building cantilever-type resonators increased their thermal sensitivity and TCF, while bringing the fabrication challenges and CMOS compatibility problem. A fixed–fixed beam-type device was built [17] to drastically increase the TCF in comparison to other well-known resonant-type devices (cantilever, tuning fork, and free-free beam). The |TCF| of 4537 ppm/K was achieved in [5], and 19,500 ppm/K was achieved in [17], with fixed–fixed beam, where CMOS allowed high density integration and low-noise readout circuit. Later, this method was improved, and a possibility of 31× improvement, with a |TCF| of 2,178,946 ppm/K in thermal sensing, was demonstrated [7] by not only combining joule heating with ambient temperature change but also keeping the pull-in voltage as small as possible. However, nonuniform temperature distribution via joule heating prevented this approach from going down to smaller scale and consequently having higher thermal sensitivity due to relatively high thermal stresses.

We present a new type of joule heating technique in this work that allows much better temperature uniformity throughout the beam in comparison to conventional joule heating. In contrast to [7], here, joule heating is applied on sidebars attached on the fixed ends of the clamped–clamped beam. This uniform temperature profile not only increases the thermal stress on the fixed ends but also allows building design with relatively higher stiffness constant, while keeping the temperature limit around 530 K [18]. This increase in return results in the increase in the sensitivity multiplier and TCF. The sensitivity multiplier was computed as 60.5×, with a 30 μm long and 0.9 μm thick device, according to both COMSOL and theory in comparison to the possibility of 31× demonstrated in [7]. The |TCF| for 60.8× improvement was calculated as 3,991,168 ppm/K in comparison to 2,178,946 ppm/K in [7], 19,500 ppm/K in [17], 4537 ppm/K in [5], 30 ppm/K in [6], 548 ppm/K in [10], 86.2 ppm/K in [19], and 29.4 ppm/K in [13]. Furthermore, this uniform heating allows the design to have the same thermal sensitivity, with a conventional joule heating scheme [7], while drastically decreasing the operation temperature and enabling long-term operation. In other words, the proposed design methodology can be used either to increase the thermal sensitivity or to decrease the operation temperature for high reliability.

Using aluminum or composite structure in building a clamped–clamped beam in a CMOS 0.6 process showed almost no difference in the sensitivity multiplier, even at different sizes, where the stiffness constant stayed the same. That's why optimum design with a critical temperature below 530 K can be designed in any fabrication process/technology (CMOS, silicon, Silicon on insulator (SOI), etc.), regardless of the material and size, as long as the stiffness constant is limited to 134 N/m.

2. Theory, Design, and Optimization

The main mechanism behind the wide-range frequency tuning is the thermal stresses created via joule heating, as demonstrated for the first time in [17]. Another study [7] conducted on CMOS clamped–clamped beam proved that thermal stress is the dominant factor on the temperature sensitivity, and the stresses caused by the increase in the pull-in force adversely affect the sensitivity. The main goal of this study is to increase the temperature uniformity throughout the beam to get further increase in the thermal stress at a relatively low temperature, and consequently get higher thermal sensitivity for microbolometer applications.

2.1. Design for Uniform Heating

One of the main disadvantages of joule heating is the well-known nonuniform temperature distribution throughout the conventional clamped–clamped beam [17] (Figure 1c). Its temperature profile can be derived by combining thermal conduction and heat generation equations:

$$T_a - T_b = \frac{I^2 \rho L^3}{2VAk} \tag{1}$$

where T_a is the maximum temperature at the center of the beam, T_b is the minimum temperature at the fixed ends, I is the current flow, L is the length, ρ is the density, V is the volume, A is the cross-sectional area, and k is the thermal conduction constant.

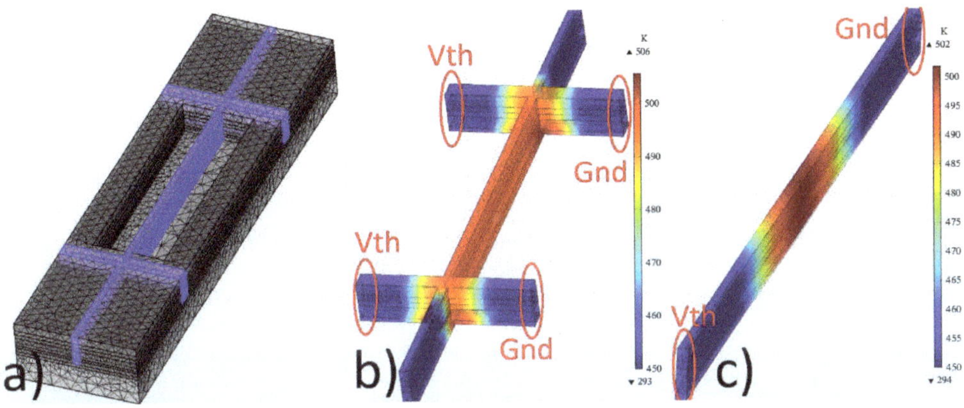

Figure 1. (**a**) UniJoule structure built via a CMOS process for performance improvement, and temperature distribution as a result of joule heating application of (**b**) UniJoule structure and (**c**) conventional clamped–clamped beam computed via COMSOL.

The conventional clamped–clamped beam was converted into a new type of structure (called UniJoule (Figure 1b)) by adding sidebars for the sake of better temperature uniformity. The design was built via a CMOS 0.6 um process, and all the details related to the fabrication process were given in [7,17], where CHF_3/O_2 was used to etch SiO_2, and XeF_2 was used to etch silicon underneath the beams. Here, the voltage (Vth) and ground (Gnd) applied on the sidebars (Figure 1b) in contrast to the conventional one, where Vth and Gnd applied on the fixed ends (Figure 1c). This enables the maximum temperature locating on fixed ends of the clamped–clamped beam embedded in the UniJoule structure (Figure 1b) and consequently creates a much more uniform temperature distribution.

The COMSOL was used to model and simulate the entire 3-layer composite beam (Figure 1a), including the substrate layer, where the first layer is SiO_2, second layer is Polysilicon, and third layer is Aluminum in the 3-layer composite beam. There are two silicon layers in the structure: the first one is the etched one to allow the beam to resonate, and the second one is the substrate that carries all the layers. The substrate thickness was kept small to decrease the simulation time, while the thermal conduction constant of the substrate was recalculated according to thermal resistance ($R = H/k$, H = substrate thickness, k = thermal conductivity) for the sake of the accuracy. The substrate bottom was kept as a fixed surface in solid mechanics, and its temperature was kept at 293 K in a heat-transfer module. The conductivity of polysilicon was set to 1.16×10^5 S/m [17], and the fine mesh with a tetrahedral structure was used to complete the UniJoule structure in COMSOL. Here, an electric current model was used to apply joule heating, a heat transfer module was used to calculate temperature distribution throughout the beam, and a solid mechanics module was used to calculate deflection, mode shapes, and resonance frequency, with respect to temperature. The high-density sweep points (up to 2.3×10^{-6} V resolution) in the application of joule heating (Vth) were applied to achieve high accuracy in the results. The convective cooling was added to the heat-transfer module for the sake of high accuracy.

2.2. Design for High Thermal Sensitivity

The detailed study was conducted on the thermal sensitivity of the UniJoule structure (Figure 2a) via COMSOL, and the results were compared to the uniform heating case (Figure 2b), where the sensitivity was derived via the use of resonance frequency, with respect to axial load [20]:

$$f = \frac{4.73^2}{2\pi L^2}\left(1 + \frac{PL^2}{EI\pi^2}\right)^{\frac{1}{2}}\left(\frac{EI}{m}\right)^{\frac{1}{2}} \tag{2}$$

$$P = \frac{4}{h}(E_1I_1 + E_2I_2)\left(\frac{(\alpha_1 - \alpha_2)T}{\frac{h}{2} + \left(\frac{2E_1I_1+E_2I_2}{hb}\right)\left(\frac{1}{E_1t_1} + \frac{1}{E_2t_2}\right)}\right) \tag{3}$$

where I is the moment of inertia, m (kg/m) is the mass per unit length, P is the total compressive axial load [21], L (m) is the length, h (m) is the width and b (m) is the thickness, and E is the Elastic modulus of the 2-layer composite beam in Equation (2). In Equation (3), t_1 is the width, α_1 is the thermal expansion constant, $E_1$is the Elastic modulus, and I_1 is the moment of inertia for the aluminum layer, t_2 is the width, α_2 is the thermal expansion constant, E_2 is the Elastic modulus, and I_2 is the moment of inertia for SiO_2 layer. Here, the 3-layer composite beam (Figure 1) was converted into an equivalent 2-layer composite beam via COMSOL, to find the total compressive axial load in Equation (3), where the first layer is SiO_2 and the second layer is Aluminum. The Equation (2) was verified and matched with the measurement results in [18], while the increase in the thermal sensitivity with the increase in the joule heating application was verified with measurements in [17,18]. Further studies [7] demonstrated that the decrease in the pull-in force around the beam-bending point increased the thermal sensitivity. However, one of the main problems and limitations for this type of device is the maximum allowable thermal stress and, consequently, the maximum allowable temperature that the resonator can tolerate. This was measured and verified around 530 K for the same type of structure in the CMOS process [18]. The possibility of a 31× improvement in thermal sensitivity was demonstrated in [7] for the conventional 57 μm long clamped–clamped beam with a maximum temperature of 530 K, around the beam bending point. In contrast to [7], here we built a UniJoule structure that allows uniform heating throughout the structure. A 60.8× ((284 + |−428|) kHz/11.7 kHz) improvement was achieved with a 30 μm long device, according to COMSOL (Figure 2a), and it was calculated around 60.5×, according to Equation (2) (Figure 2b), where the temperature (at the beam bending point [7]) was 520 K, according to COMSOL, and 529 K, according to Equation (2). This improvement is attributed to the fact that uniform heating allows for a larger stiffness constant, while keeping the critical temperature below 530 K, according to both COMSOL and Equation (2), and it consequently results in higher thermal sensitivity.

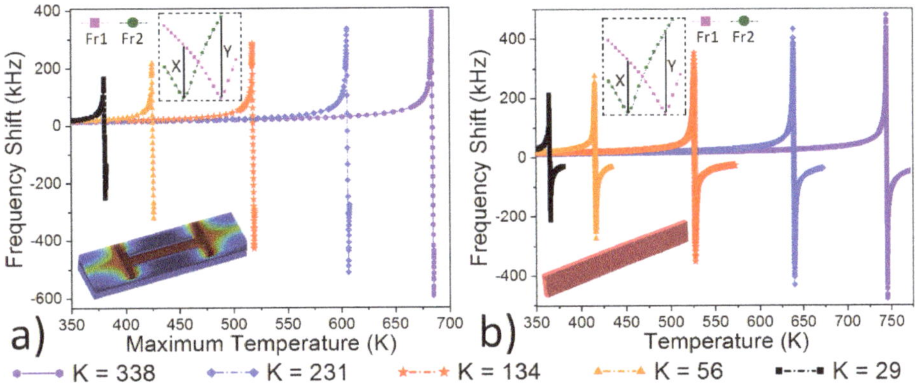

Figure 2. Frequency shift (FS = Fr1 − Fr2) with respect to 1 Kelvin change by (**a**) COMSOL for UniJoule structure with Vth application and (**b**) Equation (2) for uniformly heated conventional clamped–clamped beam, when stiffness constant (K) changed from 29 to 338 N/m; here, X represents positive max, and Y represents negative max of FS [22] in inset.

The Frequency shift (FS) was calculated by taking the difference between two resonance frequencies Fr1 and Fr2 (more details given in [7]), where the ambient temperatures were set as 293 and 294 K, respectively. The sensitivity multiplier (total maximum FS/minimum FS) was increased from 37 to 82, according to COMSOL, and from 36 to 82, according to Equation (2), when the stiffness constant increased from 29 to 338 N/m (Figure 2). The optimum structure was selected to be a 30 μm long beam, where the sensitivity multiplier is around 60.5×, with a stiffness constant of 134 N/m, according to both COMSOL and theory. A decent uniformity in temperature distribution for all UniJoule structures (Figure 2a) was achieved where the difference between the maximum and the minimum temperature throughout the beam is less than 6.5%.

More studies were conducted on the effect of the stiffness constant and material configuration on the sensitivity multiplier and maximum allowable temperature (Figure 3). The first design was built on a composite structure (Figure 1) [5], while the second one was built by only using an aluminum layer [22] in the CMOS process. The results suggest that the sensitivity multiplier depends on the stiffness constant, rather than the material type used in building the resonator. It is around 60.5×, with a temperature below 530 K, for the composite structure (according to both COMSOL and theory), and it is 60.2× at 502 K for the aluminum resonator, where the stiffness constant is 127 N/m for the aluminum design and 134 N/m for composite structure (Figure 3a). Although the sensitivity multiplier is around 60× for both aluminum and the composite structure, the aluminum resonator still shows overall better thermal sensitivity (total $|FS_{aluminum}|$ = 1306 vs. total $|FS_{composite}|$ = 712) (Figure 3a). This is attributed the fact that aluminum has a larger thermal expansion constant ($\alpha_{Aluminum} = 23.1 \times 10^{-6}$ 1/K) in comparison to composite structure, where SiO_2 ($\alpha_{SiO2} = 0.5 \times 10^{-6}$ 1/K) and Polysilicon ($\alpha_{Polysilicon} = 2.65 \times 10^{-6}$ 1/K) decreases the overall thermal expansion constant.

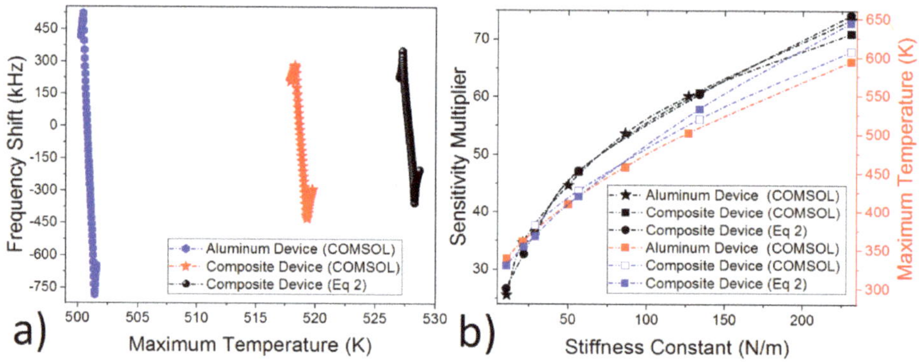

Figure 3. (**a**) Frequency shift with respect to temperature and (**b**) the relationship between stiffness constant, sensitivity multiplier, and maximum allowable temperature for both aluminum and composite structure (Figure 1) derived via COMSOL and Equation (2).

The relationship between the stiffness constant, the sensitivity multiplier, and the maximum temperature around the beam bending point was analyzed. Both the aluminum and composite structure had a relatively sharp increase in the sensitivity multiplier, where the stiffness constant was smaller than 50 (Figure 3b). The sensitivity multiplier and the maximum temperature increases with the increase in the stiffness constant. The maximum allowable temperature was set to 530 K, and that's why the maximum stiffness constant was calculated as 134 N/m, with a maximum sensitivity multiplier of 60.5×. Moreover, this method can be also used to drastically decrease the operation temperature, while keeping the sensitivity multiplier relatively high. The sensitivity multiplier around 36×, with a temperature below 383 K, and 47×, with a temperature below 428 K, can be achieved according to both COMSOL and the theory (Figures 2 and 3). This feature is especially important for long-term reliable operations.

This finding suggests that using different material or different fabrication processes in building a resonator doesn't affect the sensitivity multiplier or the maximum temperature around beam bending at all (Figure 3). The only critical parameter setting the maximum sensitivity multiplier is the stiffness constant.

2.3. *Design for Relatively Low Power Consumption*

The power consumption is another critical parameter for the sensor designs, especially with the increase in demand in the internet of things (IOT) applications. The XeF_2 process for isotropic etching of silicon layer is a well-known and widely used process, especially in CMOS–MEMS [5,17]. Here, we applied isotropic etching on the clamped–clamped beam (Figure 1b) for the sake of better thermal isolation. Design-1 has a smaller mask opening (Figure 4a inset), while design-2 has a wider mask opening (Figure 4b inset) to allow 13 µm silicon etching on both sides. This etching process decreases the thermal conduction from beam to substrate, and this allows better thermal isolation. This, in return, enables less power consumption to generate the same amount of heat on the beam, according to Equation (1). Thanks to high thermal conductivity of the silicon layer (1.3 $Wcm^{-1}{}^{\circ}C^{-1}$), this technique would allow a noticeable decrease in the power consumption.

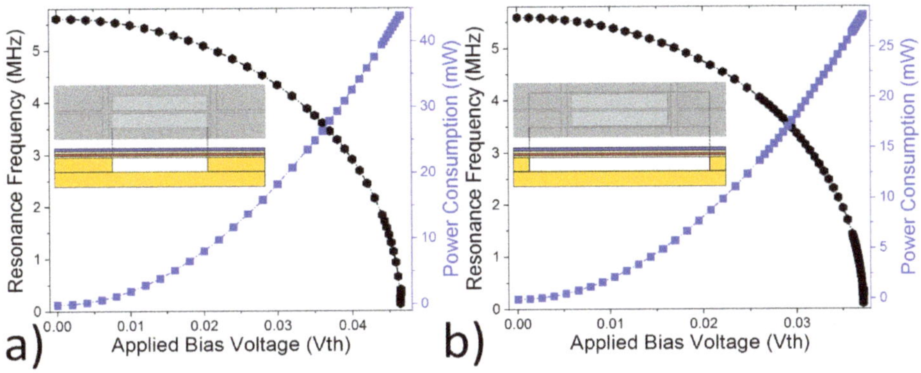

Figure 4. The power consumption and bending voltage for a UniJoule structure (Figure 1b), where (**a**) there is no isotropic on fixed-ends and (**b**) there is a 13 μm isotropic etching on both fixed-ends.

Both design-1 and design-2 were heated via applied bias voltage (Vth) (Figure 1b), till the devices reached the bending point, where the maximum thermal sensitivity was achieved (Figure 2) [7]. The bending point was reached at Vth = 0.0464 V for design-1 and Vth = 0.0372 V for design-2. This represents around 43.8 mW power consumption for design-1 and 28 mW for design-2. In other words, the power consumption was decreased by 36% via using a wider mask with the XeF_2 isotropic silicon etching process.

The effect of isotropic etching was also investigated on the thermal-sensitivity performance. Although the etching process resulted in lower power consumption, both design-1 and design-2 showed the same sensitivity multiplier. This is attributed to the fact that both designs have the same stiffness constant (Figure 3). There is only a slight difference between the two designs. Design-1 reached the bending point around 527 K, while design-2 reached it around 519 K.

3. Conclusions

A new type of joule heating scheme was demonstrated for the sake of uniform temperature distribution throughout the beam. This, in return, enabled the higher thermal stresses and higher sensitivity multiplier. Around a 60.5× improvement in thermal sensing was achieved and verified via both COMSOL and theory, while keeping the device temperature below 530 K. The very same method can also be used to drastically decrease the operation temperature and consequently enables long-term reliable operation, where 36× with a temperature below 383 K and 47× with a temperature below 428K was achieved according to both COMSOL and theory. The results suggest that the maximum sensitivity multiplier can be achieved when the stiffness constant is around 134 N/m, regardless of the materials or process used in building the devices. The |TCF| was calculated around 3,991,168 ppm/K, where the applied bias voltage (Vth) is 0.0372 V. It should be noted that this improvement (60.5×) in the thermal sensitivity was achieved without any need for a complex and expensive fabrication process or even special layers, such as 2D materials. This would be very crucial and helpful in supporting the studies conducted on medical imaging. Further improvement was achieved via isotropic etching applied on silicon layer. The power consumption decreased from 43.8 to 28 mW with the decrease in the thermal conduction according to COMSOL. This can be very beneficial for applications that requires compact size, low cost, and wireless communications, such as the internet of things (IOT).

Author Contributions: Conceptualization, H.G.; Methodology, H.G.; Software, H.G. and F.S.G.; Validation, H.G. and F.S.G.; Formal Analysis, H.G.; Investigation, H.G. and F.S.G.; Resources, H.G. and F.S.G.; Data Curation, H.G. and F.S.G.; Writing-Original Draft Preparation, H.G.; Writing-Review & Editing, H.G. and F.S.G.; Visualization, H.G. and F.S.G.; Supervision, H.G.; Project Administration, H.G.; Funding Acquisition, H.G.

Funding: This research was funded by the Scientific Research Project Foundation of Turkey (grant number 18073).

Acknowledgments: The author especially wishes to thank COMSOL for their support in setting up the simulation environment accurately for CMOS–MEMS resonator in this study.

Conflicts of Interest: The authors declare no conflicts of interest.

References

1. Marsili, F.; Verma, V.B.; Stern, J.A.; Harrington, S.; Lita, A.E.; Gerrits, T.; Vayshenker, I.; Baek, B.; Shaw, M.D.; Mirin, R.P.; et al. Detecting single infrared photons with 93% system efficiency. *Nat. Photonics* **2013**, *7*, 210–214. [CrossRef]
2. Renema, J.J.; Gaudio, R.; Wang, Q.; Zhou, Z.; Gaggero, A.; Mattioli, F.; Leoni, R.; Sahin, D.; de Dood, M.J.A.; Fiore, A.; et al. Experimental Test of Theories of the Detection Mechanism in a Nanowire Superconducting Single Photon Detector. *Phys. Rev. Lett.* **2014**, *112*, 117604. [CrossRef] [PubMed]
3. Forsberg, F.; Lapadatu, A.; Kittilsland, G.; Martinsen, S.; Roxhed, N.; Fischer, A.C.; Stemme, G.; Samel, B.; Ericsson, P.; Høivik, N.; et al. CMOS-Integrated Si/SiGe Quantum-Well Infrared Microbolometer Focal Plane Arrays Manufactured With Very Large-Scale Heterogeneous 3-D Integration. *IEEE J. Sel. Top. Quantum Electron.* **2015**, *21*, 30–40. [CrossRef]
4. Lv, J.; Que, L.; Wei, L.; Zhou, Y.; Liao, B.; Jiang, Y. Uncooled Microbolometer Infrared Focal Plane Array Without Substrate Temperature Stabilization. *IEEE Sens. J.* **2014**, *14*, 1533–1544. [CrossRef]
5. Göktaş, H.; Turner, K.L.; Zaghloul, M.E. Enhancement in CMOS-MEMS Resonator for High Sensitive Temperature Sensing. *IEEE Sens. J.* **2017**, *17*, 598–603. [CrossRef]
6. Hui, Y.; Gomez-Diaz, J.S.; Qian, Z.; Alù, A.; Rinaldi, M. Plasmonic piezoelectric nanomechanical resonator for spectrally selective infrared sensing. *Nat. Commun.* **2016**, *7*, 11249. [CrossRef]
7. Göktaş, H. Towards an Ultra-Sensitive Temperature Sensor for Uncooled Infrared Sensing in CMOS–MEMS Technology. *Micromachines* **2019**, *10*, 108. [CrossRef] [PubMed]
8. Kang, D.H.; Kim, K.W.; Lee, S.Y.; Kim, Y.H.; Keun Gil, S. Influencing factors on the pyroelectric properties of Pb (Zr, Ti) O_3 thin film for uncooled infrared detector. *Mater. Chem. Phys.* **2005**, *90*, 411–416. [CrossRef]
9. Tao, Y.; Boss, J.M.; Moores, B.A.; Degen, C.L. Single-crystal diamond nanomechanical resonators with quality factors exceeding one million. *Nat. Commun.* **2014**, *5*, 3638. [CrossRef]
10. Zhang, X.C.; Myers, E.B.; Sader, J.E.; Roukes, M.L. Nanomechanical Torsional Resonators for Frequency-Shift Infrared Thermal Sensing. *ACS Nano Lett.* **2013**, *13*, 1528–1534. [CrossRef] [PubMed]
11. Sawano, S.; Arie, T.; Akita, S. Carbon Nanotube Resonator in Liquid. *ACS Nano Lett.* **2010**, *10*, 3395–3398. [CrossRef] [PubMed]
12. Baek, I.-B.; Byun, S.; Lee, B.K.; Ryu, J.-H.; Kim, Y.; Yoon, Y.S.; Jang, W.I.; Lee, S.; Yu, H.Y. Attogram mass sensing based on silicon microbeam resonators. *Nat. Sci. Rep.* **2017**, *7*, 46660. [CrossRef] [PubMed]
13. Qian, Z.; Hui, Y.; Liu, F.; Kang, S.; Kar, S.; Rinaldi, M. Graphene–aluminum nitride NEMS resonant infrared detector. *Nat. Microsyst. Nanoeng.* **2016**, *2*, 16026. [CrossRef] [PubMed]
14. Foulgoc, B.L.; Bourouina, T.; Traon, O.L.; Bosseboeuf, A.; Marty, F.; Breluzeau, C.; Grandchamp, J.-P.; Masson, S. Highly decoupled single-crystal silicon resonators: An approach for the intrinsic quality factor. *IOP J. Micromech. Microeng.* **2006**, *16*, S45–S53. [CrossRef]
15. Laird, E.A.; Pei, F.; Tang, W.; Steele, G.A.; Kouwenhoven, L.P. A High Quality Factor Carbon Nanotube Mechanical Resonator at 39 GHz. *ACS Nano Lett.* **2012**, *12*, 193–197. [CrossRef] [PubMed]
16. Manca, N.; Pellegrino, L.; Kanki, T.; Yamasaki, S.; Tanaka, H.; Siri, A.S.; Marré, D. Programmable Mechanical Resonances in MEMS by Localized Joule Heating of Phase Change Materials. *Adv. Mater.* **2013**, *25*, 6430–6435. [CrossRef] [PubMed]
17. Göktaş, H.; Zaghloul, M.E. Tuning In-Plane Fixed–Fixed Beam Resonators with Embedded Heater in CMOS Technology. *IEEE Electron Device Lett.* **2015**, *36*, 189–191. [CrossRef]
18. Göktaş, H.; Zaghloul, M.E. The implementation of low-power and wide tuning range MEMS filters for communication applications. *Radio Sci.* **2016**, *51*, 1636–1644. [CrossRef]

19. Duraffourg, L.; Laurent, L.; Moulet, J.-S.; Arcamone, J.; Yon, J.-J. Array of Resonant Electromechanical Nanosystems: A Technological Breakthrough for Uncooled Infrared Imaging. *Micromachines* **2018**, *10*, 401. [CrossRef] [PubMed]
20. Jha, C.M. Thermal and Mechanical Isolation of Ovenized MEMS Resonator. Ph.D. Thesis, Department of Mechanical Engineering, Stanford University, Palo Alto, CA, USA, 2008.
21. Abawi, A.T. The Bending of Bonded Layers Due to Thermal Stress. Available online: http://hlsresearch.com/personnel/abawi/papers/bend.pdf (accessed on 23 October 2014).
22. Verd, J.; Uranga, A.; Abadal, G.; Teva, J.; Torres, F.; Pérez-Murano, F.; Fraxedas, J.; Esteve, J.; Barniol, N. Monolithic mass sensor fabricated using a conventional technology with attogram resolution in air conditions. *Appl. Phys. Lett.* **2007**, *91*, 013501. [CrossRef]

© 2019 by the authors. Licensee MDPI, Basel, Switzerland. This article is an open access article distributed under the terms and conditions of the Creative Commons Attribution (CC BY) license (http://creativecommons.org/licenses/by/4.0/).

micromachines

MDPI

Review

InAs/InAsSb Type-II Strained-Layer Superlattice Infrared Photodetectors

David Z. Ting *, Sir B. Rafol, Arezou Khoshakhlagh, Alexander Soibel, Sam A. Keo, Anita M. Fisher, Brian J. Pepper, Cory J. Hill and Sarath D. Gunapala

NASA Jet Propulsion Laboratory, California Institute of Technology, Pasadena, CA 91109, USA; Sir.B.Rafol@jpl.nasa.gov (S.B.R.); Arezou.Khoshakhlagh@jpl.nasa.gov (A.K.); Alexander.Soibel@jpl.nasa.gov (A.S.); Sam.A.Keo@jpl.nasa.gov (S.A.K.); Anita.M.Fisher@jpl.nasa.gov (A.M.F.); Brian.J.Pepper@jpl.nasa.gov (B.J.P.); Cory.J.Hill@jpl.nasa.gov (C.J.H.); Sarath.D.Gunapala@jpl.nasa.gov (S.D.G.)

* Correspondence: David.Z.Ting@jpl.nasa.gov

Received: 22 September 2020; Accepted: 21 October 2020; Published: 26 October 2020

Abstract: The InAs/InAsSb (Gallium-free) type-II strained-layer superlattice (T2SLS) has emerged in the last decade as a viable infrared detector material with a continuously adjustable band gap capable of accommodating detector cutoff wavelengths ranging from 4 to 15 μm and beyond. When coupled with the unipolar barrier infrared detector architecture, the InAs/InAsSb T2SLS mid-wavelength infrared (MWIR) focal plane array (FPA) has demonstrated a significantly higher operating temperature than InSb FPA, a major incumbent technology. In this brief review paper, we describe the emergence of the InAs/InAsSb T2SLS infrared photodetector technology, point out its advantages and disadvantages, and survey its recent development.

Keywords: InAs/InAsSb; type-II superlattice; infrared detector; mid-wavelength infrared (MWIR); unipolar barrier

1. Introduction

The II-VI semiconductor HgCdTe (MCT) is the most successful infrared photodetector material to date. MCT grown on nearly lattice-matched CdZnTe (CZT) substrate offers continuous cutoff wavelength (λ_{cutoff}) coverage from the short-wave infrared (SWIR) to the very long wavelength infrared (VLWIR), while providing a high quantum efficiency (QE) and low dark current for high-performance applications. In general, III-V semiconductors are more robust than their II-VI counterparts due to their stronger, less ionic chemical bonding. III-V semiconductor-based infrared focal plane arrays (FPAs) excel in operability, spatial uniformity, temporal stability, scalability, producibility, and affordability. InGaAs FPAs with $\lambda_{cutoff} \sim 1.7$ μm perform at near theoretical limit and dominate the SWIR FPA market. InSb FPAs ($\lambda_{cutoff} \sim 5.3$ μm), despite a significantly lower operating temperature than MCT, lead the mid-wavelength infrared (MWIR) market in volume due to their superior manufacturability and lower cost. The major limitation for traditional bulk III-V semiconductor detectors grown on (nearly) lattice-matched substrates is the lack of pervasive cutoff wavelength adjustability.

1.1. Advances in III-V Infrared Material

One method to achieve a wide-range cutoff wavelength adjustability in III-V semiconductors is to use bulk InAsSb alloy grown on metamorphic buffers. Recent results show that the InAsSb band gap bowing is significantly larger than previously believed, and up to $\lambda_{cutoff} \sim 12.4$ μm could be achieved [1]. In addition, nearly lattice-matched or pseudomorphic III-V semiconductor type-II superlattices (T2SLs) can provide a high degree of flexibility in cutoff wavelength. They can have a sufficient absorption strength to attain ample quantum efficiency, are less susceptible to band-to-band tunneling than bulk semiconductors [2], and are capable of achieving reduced Auger generation-recombination in properly designed structures [3].

III-V semiconductor extended SWIR (eSWIR) detectors are commonly grown on either InP or GaSb substrates. The first group includes the well-known extended InGaAs [4] and lattice-matched [5] and strain-compensated [6] InGaAs/GaAsSb type-II quantum wells. The second includes bulk GaInAsSb [7,8] and InPSb [9], as well as InAs/GaSb [10], InAs/GaSb/AlSb/GaSb [11], InAs/AlSb [12], InAs/InSb/AlSb/InSb [13], and InAs/InAsSb/AlAsSb [14] superlattices.

For MWIR and long wavelength infrared (LWIR) detectors, InAs/GaSb and InAs/InAsSb are the two most common T2SL absorbers used. The former is well-established, and has been described in detail in various review articles [15–17]. The latter [18–20] emerged more recently as an alternative with simpler growth [18], better defect tolerance, and longer minority lifetime [21], but smaller cutoff wavelength range, weaker optical absorption [22,23], and more challenging growth-direction hole transport [24,25]. In Section 3, we will discuss some basic properties of the InAs/InAsSb T2SLS in more detail.

1.2. Unipolar Infrared Detector Architecture

The unipolar barrier infrared photodetector architecture is now widely recognized as a highly effective platform for developing high-performance infrared photodetectors, as exemplified by the nBn [26], the XBn [27,28], the complementary barrier infrared detector (CBIRD) [29,30], the double heterostructure (DH) [31,32], and the pMp [33]. A unipolar barrier blocks one carrier type (electron or hole) but allows the unimpeded flow of the other. The unipolar barrier photodetector architecture can be used to lower generation-recombination (G-R) dark current by suppressing Shockley–Read–Hall (SRH) processes [26], and has also been used to reduce surface-leakage dark current [26,34,35] in devices with n-type absorbers. This has been especially beneficial for III-V semiconductor-based infrared photodiodes, many of which traditionally tend to suffer from excess depletion dark current and a lack of good surface passivation. Unipolar barrier infrared photodetectors have been successfully implemented for a variety of bulk and superlattice absorbers.

1.3. Antimonide Barrier Infrared Detectors

Taking advantage of the recent developments in the antimonide bulk and type-II superlattice infrared absorber material and the advances in the unipolar barrier infrared detector architecture, a new generation of infrared detectors has been successfully implemented in a variety of cutoff wavelengths ranging from SWIR to LWIR. In this section, we briefly mention a few examples to illustrate the effectiveness and versatility of this approach.

The 2006 paper by Maimon and Wicks on the nBn infrared detector [26] has been one of the most influential works in the field of infrared photodetectors in recent years. The performance of the nBn is enhanced by a specially constructed barrier that blocks the majority but not the minority carriers. These unipolar barriers, which are heterostructure barriers that can block one carrier type (electron or hole) but allow the un-impeded flow of the other, can be used to suppress the G-R dark current and surface leakage current [26,34,35], which are the two main dark current mechanisms that have plagued III-V semiconductor infrared detectors. The initial nBn devices used either InAs absorber grown on InAs substrate, or lattice-matched $InAs_{0.91}Sb_{0.09}$ alloy grown on GaSb substrate, with cutoff wavelengths of ~3.2 and ~4 µm, respectively. These nBn detectors could operate at much higher temperatures than InSb-based MWIR detectors, although their spectral responses do not cover the full 3 to 5 µm MWIR atmospheric transmission window like InSb detectors.

The antimonide barrier infrared detector concept has also been successfully implemented in the eSWIR using a GaInAsSb quaternary absorber and an AlGaSb or AlGaAsSb unipolar electron barrier [7,8]. By adjusting the alloy composition of an GaInAsSb infrared absorber, its cutoff wavelength can vary from ~1.8 to ~4 µm while remaining lattice-matched to the GaSb substrate. In addition, for a lattice-matched GaInAsSb absorber of a given composition, a matching AlGaAsSb electron unipolar barrier can be found for building an nBn (or XBn) detector. The left panel of Figure 1 shows an image taken with an eSWIR FPA made from such an nBn detector at the NASA Jet Propulsion Laboratory (JPL), with a cutoff wavelength of ~2.6 µm at 180 K.

The desire to develop high-performance nBn detectors with cutoff wavelength capable of covering the full 3 to 5 µm MWIR atmospheric transmission window (like InSb) motivated the exploration of T2SL nBn detectors. MWIR InAs/GaSb T2SL nBn detectors were demonstrated by Rodriguez et al. in 2007 [36], soon after the original nBn work by Maimon and Wick [26]. More recently, the InAs/InAsSb type-II strained-layer superlattice (T2SLS) has emerged as a highly versatile absorber material. Along with matching AlAsSb or AlGaAsSb electron barriers (typically lightly p-doped), the InAs/InAsSb type-II T2SLS has been used very effectively in implementing MWIR and LWIR unipolar barrier infrared detectors. Figure 1 shows images of FPAs fabricated at JPL from InAs/InAsSb T2SLS unipolar barrier infrared detectors using cutoff lengths of 5.4, 9.5, and 13.3 µm. In the next section, we will describe the InAs/InAsSb T2SLS infrared detectors in more detail.

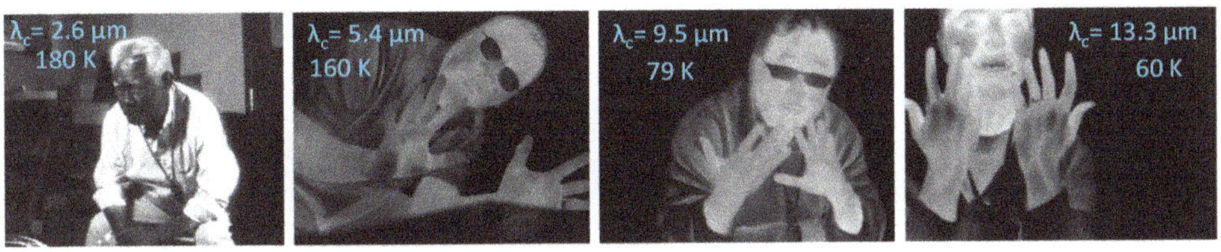

Figure 1. Images from extended-SWIR (λ_c = 2.6 µm), MWIR (λ_c = 5.4 µm), LWIR (λ_c = 9.5 µm), and VLWIR (λ_c = 13.3 µm) FPAs made from antimonide bulk and T2SL unipolar barrier infrared detectors.

2. The Emergence of InAs/InAsSb T2SLS Infrared Photodetectors

2.1. Historical Background

Though much less explored than the InAs/Ga(In)Sb T2SLS, the development of InAs/InAsSb/InSb (Gallium-free, or Ga-free) T2SLS for infrared emitter and detector applications has a long and interesting history that predates the InAs/Ga(In)Sb T2SLS detector [37]. The InAsSb/InAsSb T2SLS was originally proposed by Osbourn as a means of achieving in III-V semiconductors a smaller bandgap than bulk InAsSb in order to reach longer wavelengths [38,39]. This is made possible because of the type-II band alignment (which can lead to a superlattice band gap that is smaller than its constituent bulk materials) and strain-induced band-gap reduction. Researchers at Sandia set out to implement this idea using the InAsSb/InSb SLS grown on InSb substrate with an InSb-rich composition-graded InAsSb strain-relief buffer [40]. By 1990, S. R. Kurtz and co-workers reported a number of photodiodes and photoconductors with cutoff wavelengths ranging from 8.7 to beyond 10 µm [41–43], thus validating the concept of the LWIR InAsSb T2SLS detector. In 1995, Zhang reported continuous-wave operation 3.3–3.4 µm midinfrared lasers based on InAs/InAsSb T2SLS emitter grown on InAs substrate [44]. Meanwhile, researchers in the UK also explored the "As-rich" InAs/InAsSb T2SLS grown metamorphically on GaAs substrates. In 1995, Tang and colleagues from the Imperial College reported InAs/InAsSb T2SLSs with a strong photoluminescence (PL) intensity and infrared emission ranging from 4 to 11 µm [45]. Working with other UK research groups, it was found that As-rich InAs/InAsSb T2SLS with band gaps comparable to InSb have substantially suppressed Auger processes when compared to InSb [46]. Because of the room-temperature Auger suppression, it was suggested that this material may be attractive for mid-IR diode laser applications. In 1999, Pullin and co-workers from the Imperial College demonstrated the room-temperature operation of mid-IR light-emitting diodes based on the InAsSb T2SLS [47].

In the decade since 2000, developments in InAs/Ga(In)Sb T2SLS [2,48] flourished while research activities in the Ga-free InAsSb T2SLS were relatively dormant. Signs of renewed interest in InAsSb

T2SLS appeared in 2009, when researchers from Simon-Fraser published a paper on the growth and optical characterization of InAs/InAsSb T2SLS structures strain-balanced relative to the GaSb substrate [49]. The interest in this material as an infrared detector absorber grew stronger as the Zhang Group and collaborators reported significantly longer LWIR minority carrier lifetimes in InAs/InAsSb T2SLS than in the InAs/GaSb superlattice [21]. In 2012, the Zhang Group demonstrated an InAs/InAsSb T2SLS LWIR photodetector [19] based on the nBn device design. The Razeghi Group provided further impetus by showing the versatility of the InAs/InAsSb T2SLS, having reported LWIR [20,50,51], very long wavelength infrared (VLWIR) [52], and bias-selectable dual-band mid-wavelength infrared (MWIR)/LWIR dual-band infrared photodetectors [53]. While the Zhang Group focused on LWIR Ga-free T2SLS, the NASA Jet Propulsion Laboratory started working on Ga-free T2SLS in 2008 in an effort to develop mid-wavelength barrier infrared detectors capable of covering the full 3 to 5 μm MWIR atmospheric transmission window [54]. This effort culminated in the demonstration of InAs/InAsSb T2SLS base mid-wavelength high operating temperature barrier infrared detectors in 2010 and FPAs in 2011 [18,54], and with the subsequent development of detectors and FPAs with longer cutoff wavelengths (see Figure 1) [55].

2.2. Mid-Wavelength InAs/InAsSb T2SLS Barrier Infrared Detectors

Perhaps the most significant impact that the InAs/InAsSb T2SLS has had thus far is in the demonstration of MWIR detectors and FPAs. The MWIR FPA market is traditionally dominated in volume by InSb, with a smaller number of MCT FPAs (with higher operating temperature) filling the needs for high-performance applications. According to G. Fulop of Maxtech International, in 2018 InSb accounted for over 50% of the photodetector-based infrared FPA market (in number of units, all cutoff wavelengths). InAs/InAsSb T2SLS unipolar barrier infrared detector-based FPAs have demonstrated that, like MCT, they can operate at a much higher temperature than their InSb counterparts, all while retaining the same III-V semiconductor manufacturability advantages. The concept and initial detector and FPA results of the InAs/InAsSb T2SLS high operating temperature (HOT) barrier infrared detector (BIRD) were first described in patent documents [18]. The details were reported subsequently in the literature [56,57]; here, we briefly mention some key results.

Figure 2a shows the detector dark current density as a function of applied bias for detector temperatures ranging from 89 to 222 K for an nBn device with an InAs/InAsSb T2SL absorber and an AlAsSb electron unipolar barrier grown on GaSb substrate. Figure 2b shows the spectral quantum efficiency of the detector, demonstrating the full coverage of the MWIR transmission window. The inset shows the dependence of the detector 50%-peak-QE cutoff wavelength as a function of temperature. At 150 K, the 50% cutoff wavelength is 5.37 μm, and the quantum efficiency at 4.5 μm is ~52% without anti-reflection coating. In general, the dark current performance is quite good. Under a −0.2 V bias, the dark current density at 157K is 9.6×10^{-5} A/cm^2, which a factor ~4.5 higher than that predicted by the MCT Rule 07 [58] for a cutoff wavelength of 5.4 μm. At 150K, the dark-current-limited and the f/2 black-body (300 K background in 3–5 μm band)-specific detectivities are, respectively, 4.6×10^{11} and 3.0×10^{11} cm-Hz$^{1/2}$/W.

The detector material was used to fabricate 24-μm pitch, 640 × 512 format arrays hybridized to the SBF-193 readout integrated circuit (ROIC). The cutoff wavelength is ~5.4 μm, closely matching that of the InSb FPA. A 160 K two-point corrected image taken with a resulting FPA is shown in Figure 1. At 160 K, the 300 K background, f/2 aperture mean noise equivalent differential temperature (NEDT) is 18.7 mK, with a standard deviation of 9.2 mK and a NEDT operability of 99.7%. The ion-implanted planar InSb FPAs and molecular beam epitaxy (MBE)-grown epi-InSb FPAs typically operate at 80 K and 95–100 K [59], respectively. The InAs/InAsSb T2SLS-based mid-wavelength high operating temperature barrier infrared detector (HOT-BIRD) FPA has demonstrated significant operating temperature advantages over InSb. For 300 K background imaging applications, the mid-wavelength HOT-BIRD essentially combines the higher operating advantage of MCT with the III-V semiconductor material robustness advantages of InSb, thus firmly establishing the InAs/InAsSb T2SLS as a viable infrared FPA technology.

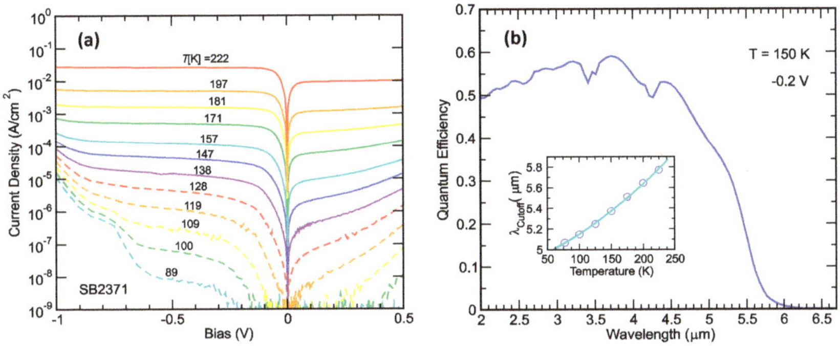

Figure 2. (**a**) Dark current density–voltage characteristics for an InAs/InAsSb type-II strained-layer superlattice nBn detector taken at temperatures ranging from 89 to 222 K. (**b**) Back-side illuminated spectral quantum efficiency (QE) for the same detector without anti-reflection coating, taken under a −0.2 V bias at 150 K. The inset shows the 50% peak QE cutoff wavelength as a function of temperature.

3. Basic Properties of the InAs/InAsSb T2SLS

In this section, we describe some basic properties of the InAs/InAsSb T2SLS and discuss its advantages and disadvantages as an infrared detector absorber material.

3.1. InAs/InAsSb T2SLS Electronic Properties

Figure 3 shows the energy band diagrams for two (m,n)-InAs/InAs$_{0.5}$Sb$_{0.5}$ superlattices (m and n, respectively, being the number of monolayers of InAs and InAsSb in each superlattice period) that are strain-balanced with respect to the GaSb substrate. For a given layer thickness ratio (m/n), the InAsSb alloy composition is selected to achieve strain balancing. Varying the superlattice period ($m + n$) changes the band gap and the corresponding cutoff wavelength ($\lambda_{cutoff} = hc/E_g$). In such a strain-balanced superlattice, typically the InAs layer is under slight tensile strain while InAsSb is under a relatively high compressive strain. Therefore, a comparatively thick InAs layer is required for strain balance against the thinner InAsSb layer. The InAs and InAsSb layers are, respectively, electron and hole quantum wells. Because the InAsSb electron barriers separating the InAs electron quantum wells are relatively weak, the c1 (lowest superlattice conduction band state) wavefunction is only weakly confined, as can be seen in the c1 probability density plots in Figure 3. On the other hand, it can also be seen in Figure 3 that the hh1 (heavy-hole 1, the highest superlattice valence band state) wavefunction is substantially localized in the InAsSb valence band quantum wells, which are separated by relatively thick layers of InAs hole barriers. This is reflected in the corresponding band structure, which is discussed next.

Figure 4 shows the energy band structures for the same two superlattices shown in Figure 3: (a) MWIR superlattice (16,4)-InAs/InAs$_{0.5}$Sb$_{0.5}$ with $E_g = 0.217$ eV and $\lambda_{cutoff} = 5.7$ μm, and (b) LWIR superlattice (28,7)-InAs/InAs$_{0.5}$Sb$_{0.5}$ with $E_g = 0.116$ eV and $\lambda_{cutoff} = 10.7$ μm. The relatively weak c1 state confinement and the stronger hh1 state confinement are reflected in the band structure, with the c1 subband having a strong dispersion along the growth direction, while the hh1 subband is nearly dispersionless along the growth direction. This indicates that, along the growth direction, electron transport is more favorable than hole transport (even more so than in the typical bulk semiconductors). The splitting between the hh1 and lh1 (light-hole 1) band is favorable for suppressing band-to-band tunneling, which depends on the c1-lh1 gap [2,37]. For the LWIR superlattice in Figure 4b, the hh1-lh1 splitting is actually larger than the c1-hh1 superlattice band gap. This is favorable for Auger-7 suppression [3]. For a more detailed discussion on the effect of superlattice band structure on infrared absorbers, see, for example, Reference [17].

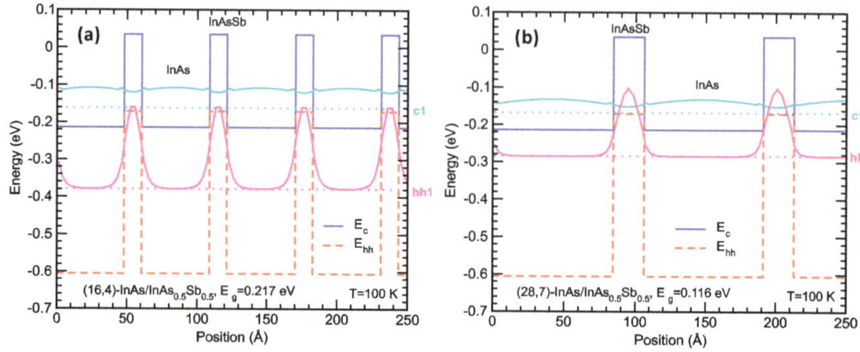

Figure 3. Energy band diagrams showing the bulk InAs and InAsSb conduction band edge (Ec) and valence band edge (E_{hh}), the superlattice zone-center c1 and hh1 energy levels (dotted lines), and the corresponding c1 and hh1 state probability densities (solid lines) for (**a**) the (16,4)-InAs/$InAs_{0.5}Sb_{0.5}$ strained-layer superlattice (SLS), and (**b**) the (28,7)-InAs/$InAs_{0.5}Sb_{0.5}$ SLS at 100 K. The InAs and InAsSb layers are, respectively, under tensile and compressive strain on the GaSb substrate.

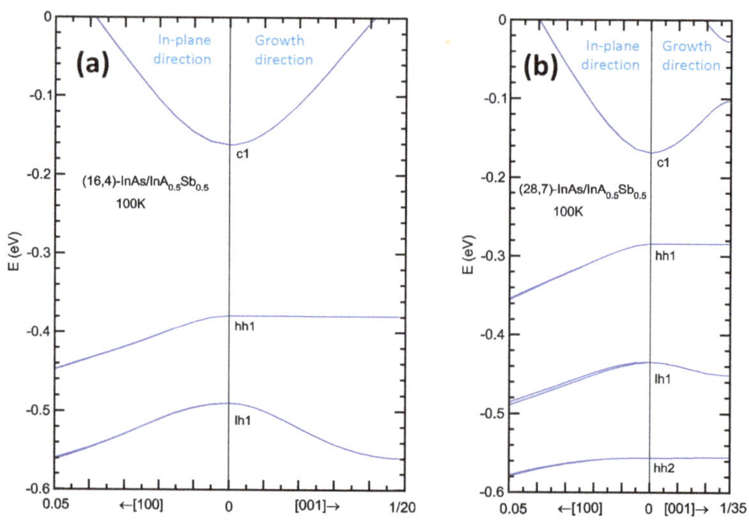

Figure 4. Energy band structure along an in-plane direction, and along the growth direction for (**a**) the (16,4)-InAs/$InAs_{0.5}Sb_{0.5}$ strained-layer superlattice (SLS) (Eg = 0.217 eV; λ_{cutoff} = 5.7 μm), and (**b**) the (28,7)-InAs/$InAs_{0.5}Sb_{0.5}$ SLS (Eg = 0.116 eV; λ_{cutoff} = 10.7 μm), on GaSb substrate at 100 K.

Figure 5a shows how the calculated c2, c1, hh1, lh1, and hh2 energy levels vary in strain-balanced (m,n)-InAs/$InAs_{0.5}Sb_{0.5}$ superlattices, $m/n = 4$, as functions of P, where $P = (m + n)$ is the superlattice period in units of monolayers. To help visualize the location of the superlattice energy levels, the background of Figure 5 shows a superlattice energy band diagram with the InAs and InAsSb strained conduction, heavy-hole, and light-hole band edges. Note that, since all the superlattices here have the same (m/n) ratio and therefore have the same energy band diagram except for a horizontal-axis scaling factor, the background band diagram in Figure 5 is shown with an arbitrary horizontal distance scale and can therefore be shared by all the calculated structures. As the superlattice period increases, the hh1 level rises while the c1 level stays relatively constant at just a few tens of meV above the InAs conduction band edge. The relatively constant c1 level can be exploited for constructing heterostructures with aligned conduction bands. For instance, an MWIR superlattice can be used as a unipolar hole barrier for an LWIR superlattice. As the superlattice period increases, the (c1-hh1) band gap decreases while the hh1-lh1 splitting increases. As mentioned earlier, a hh1-lh1 splitting larger than the c1-hh1 band gap is favorable for Auger-7 suppression. Figure 5b replots the c1 and hh1 levels, now as functions of the cutoff wavelength calculated from the superlattice band gap. To this, we add also the results for a set of strain-balanced (m,n)-InAs/$InAs_{0.6}Sb_{0.4}$ superlattices with $(m/n) = 3$, and a set of $(m,7)$-InAs/GaSb superlattices. The c1 levels for the two sets of InAs/InAsSb superlattices are both approximately independent of the cutoff wavelength and have approximately the same value. For the $(m,7)$-InAs/GaSb superlattices, the valence band edge remains approximately constant, since we fix the GaSb hole quantum well width at seven monolayers. In general, the conduction band edges of the InAs/InAsSb superlattices are low compared to those for the InAs/GaSb superlattices.

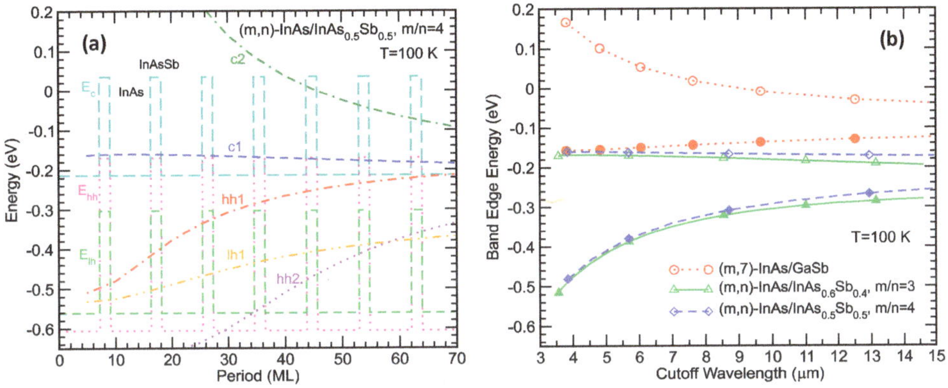

Figure 5. (**a**) The calculated c2, c1, hh1, lh1, and hh2 energy levels for a set of $(4n,n)$-InAs/$InAs_{0.5}Sb_{0.5}$ superlattices, plotted as functions of superlattice period in monolayers (MLs). The background shows a relevant superlattice energy band diagram, with an arbitrary horizontal length scale. (**b**) The calculated superlattice conduction band edge (E_{c1}, open symbols) and valence band edge (E_{hh1}, solid symbols) as functions of the cutoff wavelength for sets of $(m,7)$-InAs/GaSb, $(3n,n)$-InAs/$InAs_{0.6}Sb_{0.4}$ and $(4n,n)$-InAs/$InAs_{0.5}Sb_{0.5}$ superlattices.

3.2. InAs/InAsSb T2SLS Advantages

The InAs/InAsSb T2SLS has some advantages over the more established InAs/GaSb type-II superlattice (T2SL). The InAs/InAsSb T2SLS is easier to grow, has longer minority carrier lifetimes, and appears to have a better defect tolerance. Figure 6 illustrates the molecular beam epitaxy (MBE) shutter sequence used in the growths of InAs/GaSb T2SL and InAs/InAsSb T2SLS. In principle, the growth of the InAs/InAsSb T2SLS involves only opening and closing the Sb shutter, while the In and As shutters can stay open throughout [18]. This compares to the need to the use of four shutters in the InAs/GaSb T2SL. The growth of the InAs/GaSb T2SL is actually considerably more complicated than indicated in the simplified illustration of Figure 6, which does not include the strain-balancing interfaces required to achieve a high material quality. Thus, in general, the InAs/InAsSb T2SLS is simpler to grow.

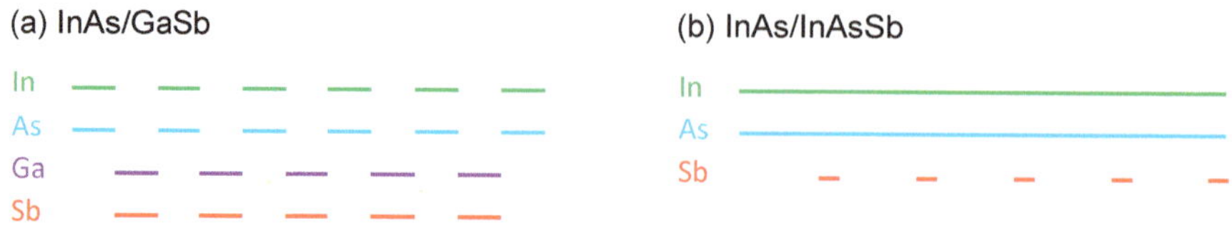

Figure 6. Schematic illustration of mechanical shutter sequences used in growing (**a**) InAs/GaSb and (**b**) InAs/InAsSb superlattices.

InAs/InAsSb T2SLS has demonstrated longer minority carrier lifetimes than the InAs/GaSb T2SL [21,60,61]. For instance, the MW InAs/GaSb T2SL minority carrier lifetime has been reported at ~80 ns [62], while non-intentionally doped MW InAs/InAsSb T2SLS has exhibited minority carrier lifetime values ranging from 1.8 [61] to 9 μs [60], with a Shockley–Read–Hall (SRH) lifetime of ~10 μs [61].

There has also been evidence suggesting that the longer minority carrier lifetime of InAs/InAsSb T2SLS is related to its defect tolerant. Tang et al. pointed out in their 1995 work [45] that, despite the high threading dislocations expected in the InAs/InAsSb SLSs grown with metamorphic buffers on highly lattice-mismatched GaAs substrates, the Shockley–Read contributions to recombination rates were low, as indicated by the strong photoluminescence (PL) intensity observed. It was hypothesized that this is due to the fact that defect state energy levels in the InAs/InAsSb T2SLS are resonant with the conduction band, rather than in the band gap where they could contribute to carrier recombination. The idea that the defect energy levels are in the conduction band was confirmed in recent pressure-dependence PL experiments on an MWIR InAs/InAsSb T2SLS grown on GaSb [63]. The reason for this is that the InAs/InAsSb T2SLS conduction band edge, like that for bulk InAs, is low, as can be seen in the theoretical results in Figure 5b. At JPL, we have seen anecdotal evidence for defect tolerance. One of the earliest MWIR InAs/InAsSb T2SLS nBn detector wafers was grown on a vintage 2011 developmental 4-inch-diameter GaSb substrate. At that time, the 4-inch substrate surface polishing was not nearly as mature as it is today. The fact that we were nevertheless able to obtain reasonable FPA results can be attributed in part to the defect tolerance of the InAs/InAsSb superlattice (the nBn device architecture being another major contributing factor) [54].

3.3. InAs/InAsSb T2SLS Challenges

The disadvantages of the InAs/InAsSb T2SLS compared to the InAs/Ga(In)Sb T2SLS are (1) weaker LWIR absorption, and (2) more challenging LWIR hole transport. Both are the results of the fact that for the InAs/InAsSb T2SLS a longer superlattice period is required in order to reach the same LWIR band gap as the InAs/Ga(In)Sb T2SLS.

Figure 7 shows the calculated cutoff wavelength as a function of superlattice period for the same three sets of superlattices discussed in Figure 5b. The cutoff wavelength is derived from the calculated superlattice band gap using the relationship λ_{cutoff} [μm] = $1.24/E_g$ [eV]. In the MWIR range, the three set of superlattices have comparable periods. As the cutoff wavelength increases, the periodicity advantage of the InAs/GaSb T2SL becomes more pronounced. Comparison between the two set of (m,n)-InAs/InAsSb superlattices show that the set with higher (m/n) ratio and higher Sb fraction InAsSb alloy is more favorable. In a type-II superlattice, the band-edge electron and hole wavefunctions are localized in different layers (see Figure 3). A longer superlattice period reduces the electron-hole wavefunction overlap, leading to weaker oscillator strength and smaller absorption coefficient. Early theoretical analysis by Grein et al. [64] showed that, compared to the InAs/GaInSb superlattice, the InAs/InAsSb superlattice requires wider InAs layers to achieve a comparable band gap and therefore produces smaller optical matrix elements; the calculated absorption coefficients for a 11 μm cutoff InAs/GaInSb T2SLS and a 10 μm cutoff InAs/InAsSb T2SLS are 2000 and 1500 cm^{-1}, respectively. More recently, Vurgaftman et al. calculated the absorption coefficients for LWIR superlattices with band gaps of ~0.1 eV (corresponding to cutoff wavelengths of $\lambda_{cutoff} = 10–12$ μm), and showed that the InAs/InAsSb T2SLS absorption coefficient is approximately half as large as that for the InAs/GaSb T2SL [23]; at λ = 8 μm, the absorption coefficients are ~1250 and ~700 cm^{-1} for the InAs/GaSb (70 Å period) and the InAs/InAsSb (125 Å period) superlattices, respectively. Klipstein et al. modeled the dependence of the LWIR superlattice detector spectral quantum efficiency (QE) on the diffusion length, and concluded that, even for a very large hypothetical diffusion length, the InAs/InAsSb T2SLS has a significantly lower QE than the InAs/GaSb T2SL because of its weaker absorption coefficient [22].

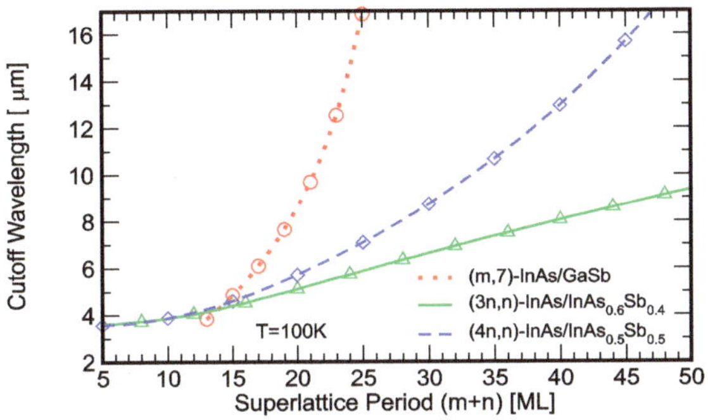

Figure 7. Calculated cutoff wavelength for the sets of (*m*,7)-InAs/GaSb, (3*n*,*n*)-InAs/$InAs_{0.6}Sb_{0.4}$, and (4*n*,*n*)-InAs/$InAs_{0.5}Sb_{0.5}$ superlattices, as functions of the superlattice period in monolayers (MLs).

The longer period of the LWIR InAs/InAsSb T2SLS also results in larger growth-direction hole conductivity effective masses, which in turn lead to lower vertical hole mobility and shorter diffusion length. From the textbook expressions for diffusion length ($L = \sqrt{D\tau_r}$), diffusivity ($D = \mu k_B T/e$), and mobility ($\mu = e\tau_c/m^{**}$), we see that the diffusion length depends explicitly on the conductivity effective mass through the expression $L_i = \left[\left(k_B T/m_i^{**}\right)\tau_r \tau_{c,i}\right]^{1/2}$, where m_i^{**} is the conductivity effective mass, τ_r is the minority carrier recombination lifetime, $\tau_{c,i}$ is the collision (momentum relaxation) time, and i is the direction index. It can be seen that a large growth-direction conductivity effective mass reduces diffusion length, which in turn limits the practical absorber thickness. In the case of LWIR InAs/InAsSb T2SLS, which has s weaker absorption coefficient than the corresponding InAs/GaSb T2SL and bulk absorbers, the inadequate absorber thickness limits the attainable quantum efficiency. Calculations by Klipstein et al. demonstrate from another perspective the strong QE dependence on the hole diffusion length in an LWIR InAs/InAsSb T2SLS XBn detector: for a 9.7 μm cutoff detector with a fixed 5 μm thick n-type absorber, the QE at 8.5 μm is 40% for a 5 μm hole diffusion length but drops to only 10% for a 1 μm hole diffusion length [22].

Figure 8 shows the calculated electron and hole conductivity effective masses along the growth direction ($m_{n,z}^{**}$ and $m_{p,z}^{**}$, respectively) as functions of the cutoff wavelengths for the same sets of superlattices discussed in Figure 7. The conductivity effective masses are thermally averaged quantities which take into consideration the anisotropy and non-parabolicity in the superlattice band structure; detail discussions can be found in References [24,25]. Figure 8a shows the calculated growth-direction electron conductivity effective mass $m_{n,z}^{**}$. For the InAs/GaSb and the $(m/n) = 4$ InAs/$InAs_{0.5}Sb_{0.5}$ superlattices, $m_{n,z}^{**}$ values are quite small since the c1 wavefunctions are only weakly confined in the relative shallow conduction band quantum wells separated by thin InAsSb barriers (see Figure 3). For $(m/n) = 3$ InAs/$InAs_{0.6}Sb_{0.4}$ superlattices (with lower Sb fraction InAsSb), the electron effective mass $m_{n,z}^{**}$ can increase with the cutoff wavelength to rather large values because of the much longer superlattice period; see Table 1 for a comparison of the superlattice period and $m_{n,z}^{**}$ for three different superlattices, all with cutoff wavelengths in the 12–13 μm range. Figure 8b shows the calculated growth-direction hole conductivity effective mass $m_{p,z}^{**}$ for the same three sets of superlattices. Again, the $m_{p,z}^{**}$ values for the three sets of superlattices are very similar in the MWIR. However, as the cutoff wavelength increases, the superlattice periods required to reach the same cutoff wavelength diverge (see Figure 7), and the $m_{p,z}^{**}$ for the three sets of superlattices also diverge significantly (see Table 1 for specific examples). For the InAs/InAsSb superlattices, $m_{p,z}^{**}$ can be very large when the InAsSb hole quantum wells are separated by a wider InAs layer (see Figure 3). Here, the higher Sb fraction in the InAsSb alloy decreases the InAs/InAsSb T2SLS period, and is therefore especially helpful in reducing the growth-direction hole conductivity effective masses. Since the growth-direction electron conductivity effective masses $m_{n,z}^{**}$ are much smaller than the hole effective masses $m_{p,z}^{**}$, the diffusion length can be much longer in the p-type InAs/InAsSb T2SLS than in the n-type, which would be more favorable for achieving a higher quantum efficiency. However, for reticulated detector structures with p-type InAs/InAsSb T2SLS absorbers, the exposed side-wall surfaces are inverted to degenerate the n-type (like in InAs) and passivation is often needed for reducing the surface related dark currents; for a discussion on surface leakage dark current mechanisms in unipolar barrier detectors, see Reference [34,65–67]. Despite the challenges, (V)LWIR n-type and p-type InAs/InAsSb T2SLS detectors with cutoff wavelengths ranging from 8 to 15 μm and front-side illuminated quantum efficiencies from 2.5% to 40% have been reported [19,20,50–52,68,69].

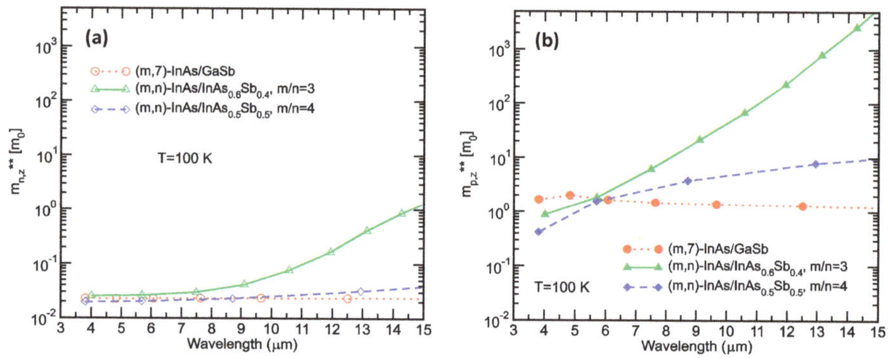

Figure 8. Growth-direction (**a**) electron and (**b**) hole conductivity effective masses for two families of InAs/InAsSb superlattices, and a set of InAs/GaSb superlattices.

Table 1. Growth direction electron and hole conductivity effective masses in units of bare electron mass for three superlattices with approximately the same cutoff wavelength.

Superlattice	Period ($m + n$) [monolayer]	λ_{cutoff} [μm]	$m_{n,z}^{**}$ [m_0]	$m_{p,z}^{**}$ [m_0]
(16,7)-InAs/GaSb	23	12.5	0.0233	1.29
(54,18)-InAs/$InAs_{0.6}Sb_{0.4}$	72	12.0	0.166	222
(32,8)-InAs/$InAs_{0.5}Sb_{0.5}$	40	13.0	0.0313	7.89

3.4. Concepts for Addressing LWIR InAs/InAsSb T2SLS Challenges

LWIR and VLWIR InAs/InAsSb T2SLS have relatively long superlattice periods and therefore have relatively weak optical absorptions and short hole diffusion lengths, which are challenging for achieving a high quantum efficiency. We recently explored theoretically some ideas for addressing the challenges for the LWIR InAs/InAsSb superlattices [70]; here, we briefly summarize the results. In comparing the $(m/n) = 3$ InAs/$InAs_{0.6}Sb_{0.4}$ and the $(m/n) = 4$ InAs/$InAs_{0.5}Sb_{0.5}$ superlattices in the discussions above, we found that increasing the Sb fraction in the InAsSb alloy can reduce the InAs/InAsSb superlattice period significantly. In fact, at sufficiently high Sb fraction (~75%), InAs/InAsSb can match InAs/GaSb in terms of the superlattice period required to reach a given cutoff wavelength [70]. However, high Sb fraction InAs/InAsSb superlattices are more prone to Sb segregation [71–74], which can negate the period-reduction benefit of high-fraction Sb. Polytype superlattices [75] such as the "W" [31], "M" [76], and "N" [77] structures have been used for improving the oscillator strength over the basic InAs/GaSb

T2SL. The analogous polytype W, M, and N superlattices formed by inserting thin AlAsSb barrier layers in InAs/InAsSb T2SLS have been considered as a means for increasing electron-hole wavefunction overlap for stronger optical absorption. This strategy turns out to be unfavorable because the presence of the AlAsSb barriers leads to increased band gap, and therefore increases the superlattice period required to reach a given cutoff wavelength [70]. Metamorphic growth on virtual substrates with larger lattice constants than GaSb can decrease the superlattice period needed to reach a specified cutoff wavelength, but this benefit should be weighed against the need for metamorphic buffer growth and the resulting higher defect density [70]. Finally, as mentioned previously, p-type InAs/InAsSb T2SLS has a longer diffusion length than n-type, but reticulated detector pixels with exposed p-type absorber side-wall surface would require passivation to suppress the surface leakage dark current.

4. Recent Development and Outlook

InAs/InAsSb T2SLS infrared detectors have been under very active development in recent years, with reported results by research groups worldwide in MWIR [18,56,57,78–82], LWIR [19,20,50,51,83,84], and VLWIR [52,85] detectors, as well as in MWIR/LWIR [53] and LWIR/VLWIR [68] bias-switchable dual-band detectors. Microlens [85] and resonant cavity structures [86] have been used to enhance the photo-response of InAs/InAsSb T2SLS infrared detectors. Studies of transport properties [87–90] and defect levels [63] have led to an improved understanding of InAs/InAsSb superlattices. While most of the more recent InAs/InAsSb T2SLS structures have been grown on (100) GaSb substrates by MBE, growths by a variety of modes have also been demonstrated. MWIR and LWIR detectors have been grown by metal organic chemical vapor deposition (MOCVD) [91–94]. Growths on GaAs [84], Si [80,95], Ge-Si [96], and AlSb (via metamorphic buffer on GaSb) [97] substrates have been reported. The growth of MWIR and LWIR detectors on (211)A and B, and (311)A and B GaSb substrates have also been demonstrated [69,98].

The MWIR InAs/InAsSb T2SLS FPA rivals InSb in manufacturability and affordability, but offers a 40 to 50 K higher operating temperature advantage, which can lead to a lower cryocooler size, weight, and power (SWaP). As such, it is poised to replace the InSb FPA, a major incumbent technology, in many imaging applications. The MWIR InAs/InAsSb T2SLS detectors and FPAs have also demonstrated very low dark current densities at lower temperatures [99], and are therefore suitable for more demanding applications such as the CubeSat-based hyperspectral imaging of 300 K scenes while operating in an intermediate temperature range (100–120 K) [100]. Although LWIR InAs/InAsSb T2SLS FPAs have only achieved a moderate quantum efficiency, their demonstrated large-format capability and high uniformity and operability makes them already suitable for applications such as LWIR imaging for Earth remote sensing applications, where photon flux is abundant. Dual-band FPAs are also expected to find applications because of the manufacturability and cost effectiveness of InAs/InAsSb T2SLS FPAs. The further development of InAs/InAsSb T2SLS infrared detectors will continue to benefit from the infrastructure established largely during the the VISTA Program [101,102], including the availability of large-diameter format GaSb substrates [103–106] and multi-wafer growth capability at commercial foundries [107,108].

Funding: The research was carried out at the Jet Propulsion Laboratory, California Institute of Technology, under a contract with the National Aeronautics and Space Administration (80NM0018D0004).

Acknowledgments: The authors would like to thank S. Bandara, A.J. Ciani, R.E. DeWames, D. Forrai, C.H. Grein, L. Höglund, M.A. Kinch, P.C. Klipstein, A.W.K. Liu, D. Lubyshev, S. Maimon, T.S. Pagano, P. Pinsukanjana, D.R. Rhiger, J.N. Schulman, W.E. Tennant, M.Z. Tidrow, Y. Wei, G.W. Wicks, R.Q. Yang, and Y.-H. Zhang for helpful discussions.

Conflicts of Interest: The authors declare no conflict of interest.

References

1. Svensson, S.P.; Sarney, W.L.; Hier, H.; Lin, Y.; Wang, D.; Donetsky, D.; Shterengas, L.; Kipshidze, G.; Belenky, G. Band gap of InAs1−xSbx with native lattice constant. *Phys. Rev. B* **2012**, *86*, 245205. [CrossRef]
2. Smith, D.L.; McGill, T.C.; Schulman, J.N. Advantages of the HgTe-CdTe superlattice as an infrared detector material. *Appl. Phys. Lett.* **1983**, *43*, 180–182. [CrossRef]
3. Grein, C.H.; Young, P.M.; Ehrenreich, H. Minority carrier lifetimes in ideal InGaSb/InAs superlattices. *Appl. Phys. Lett.* **1992**, *61*, 2905. [CrossRef]
4. Besikci, C. Extended short wavelength infrared FPA technology: Status and trends. In Proceedings of the Quantum Sensing and Nano Electronics and Photonics XV, San Francisco, CA, USA, 27 January–1 February 2018.
5. Sidhu, R.; Duan, N.; Campbell, J.C.; Holmes, A.L. A Long-Wavelength Photodiode on InP Using Lattice-Matched GaInAs–GaAsSb Type-II Quantum Wells. *IEEE Photon. Technol. Lett.* **2005**, *17*, 2715. [CrossRef]
6. Chen, B.; Holmes, A.L., Jr. InP-based short-wave infrared and midwave infrared photodiodes using a novel type-II strain-compensated quantum well absorption region. *Opt. Lett.* **2013**, *38*, 2750. [CrossRef]
7. Craig, A.; Jain, M.; Wicks, G.; Golding, T.; Hossin, K.; McEsan, K.; Howle, C.; Percy, B.; Marshall, A. Short-wave infrared barriode detectors using InGaAsSb absorption material lattice matched to GaSb. *Appl. Phys. Lett.* **2015**, *106*, 201103. [CrossRef]
8. Ting, D.Z.; Hill, C.J.; Soibel, A.; Bandara, S.V.; Gunapala, S.D. High Operating Temperature Barrier Infrared Detector with Tailorable Cutoff Wavelength. U.S. Patent Application No. 2010/0072514 A1, 6 January 2015.
9. Cohen-Elias, D.; Uliel, Y.; Cohen, N.; Shafir, I.; Westreich, O.; Katz, M. Short wavelength infrared pBn GaSb/AlAsSb/InPSb photodetector. *Infrared Phys. Technol.* **2017**, *85*, 81. [CrossRef]
10. Huang, J.; Ma, W.; Wei, Y.; Zhang, Y.; Cui, K.; Cao, Y.; Guo, X.; Shao, J. How to use type II InAs/GaSb superlattice structure to reach detection wavelength of 2–3 μm. *IEEE J. Quant. Electron.* **2012**, *48*, 1322. [CrossRef]
11. Haddadi, A.; Chevallier, R.; Dehzangi, A.; Razeghi, M. Extended short-wavelength infrared nBn photodetectors based on type-II InAs/AlSb/GaSb superlattices with an AlAsSb/GaSb superlattice barrier. *Appl. Phys. Lett.* **2017**, *110*, 101104. [CrossRef]
12. Cohen-Elias, D.; Snapi, N.; Klin, O.; Weiss, E.; Shusterman, S.; Meir, T.; Katz, M. Minority carrier diffusion length for electrons in an extended SWIR InAs/AlSb type-II superlattice photodiode. *Appl. Phys. Lett.* **2017**, *111*, 201106. [CrossRef]
13. Cohen-Elias, D.; Uliel, Y.; Klin, O.; Snapi, N.; Weiss, E.; Shafir, I.; Westreich, O.; Katz, M. Short wavelength infrared InAs/InSb/AlSb type-II superlattice photodetector. *Infrared Phys. Technol.* **2017**, *84*, 82. [CrossRef]
14. Haddadi, A.; Suo, X.V.; Adhikary, S.; Dianat, P.; Chevallier, R.; Hoang, A.M.; Razeghi, M. High-performance short-wavelength infrared photodetectors based on type-II InAs/InAs1-xSbx/AlAs1-xSbx superlattices. *Appl. Phys. Lett.* **2015**, *107*, 141104. [CrossRef]
15. Bürkle, L.; Fuchs, F. InAs/(GaIn)Sb superlattices: A promising material system for infrared detection. In *Handbook of Infrared Detection Technologies*; Henini, M., Razeghi, M., Eds.; Elsevier Science: Oxford, UK, 2002; pp. 159–189.
16. Razeghi, M.; Mohseni, F. GaSb/InAs superlattices for infrared FPAs. In *Handbook of Infrared Detection Technologies*; Henini, M., Razeghi, M., Eds.; Elsevier Science: Oxford, UK, 2002; pp. 191–232.
17. Ting, D.Z.-Y.; Soibel, A.; Höglund, L.; Nguyen, J.; Hill, C.J.; Khoshakhlagh, A.; Gunapala, S.D. Type-II Superlattice Infrared Detectors. In *Semiconductors and Semimetals*; Advances in Infrared Photodetectors; Gunapala, S., Rhiger, D., Jagadish, C., Eds.; Elsevier Academic Press: Cambridge, MA, USA, 2011; Volume 82, pp. 1–57.
18. Ting, D.Z.; Khoshakhlagh, A.; Soibel, A.; Hill, C.J.; Gunapala, S.D. Barrier Infrared Detector. U.S. Patent Application No. 13/197,588, 2011.
19. Kim, H.S.; Cellek, O.O.; Lin, Z.-Y.; He, Z.-Y.; Zhao, X.-H.; Liu, S.; Li, H.; Zhang, Y.-H. Long-wave infrared nBn photodetectors based on InAs/InAsSb type-II superlattices. *Appl. Phys. Lett.* **2012**, *101*, 161114. [CrossRef]
20. Haddadi, A.; Chen, G.; Chevallier, R.; Hoang, A.M.; Razeghi, M. InAs/InAs1−xSbx type-II superlattices for high performance long wavelength infrared detection. *Appl. Phys. Lett.* **2014**, *105*, 121104. [CrossRef]

21. Steenbergen, E.H.; Connelly, B.C.; Metcalfe, G.D.; Shen, H.; Wraback, M.; Lubyshev, D.; Qiu, Y.; Fastenau, J.M.; Liu, A.W.K.; Elhamri, S.; et al. Significantly improved minority carrier lifetime observed in a long-wavelength infrared III-V type-II superlattice comprised of InAs/InAsSb. *Appl. Phys. Lett.* **2011**, *99*, 251110. [CrossRef]
22. Klipstein, P.C.; Livneh, Y.; Glozman, A.; Grossman, S.; Klin, O.; Snapi, N.; Weiss, E. Modeling InAs/GaSb and InAs/InAsSb Superlattice Infrared Detectors. *J. Electron. Mater.* **2014**, *43*, 2984. [CrossRef]
23. Vurgaftman, I.; Belenky, G.; Lin, Y.; Donetsky, D.; Shterengas, L.; Kipshidze, G.; Sarney, W.L.; Svensson, S.P. Interband absorption strength in long-wave infrared type-II superlattices with small and large superlattice periods compared to bulk materials. *Appl. Phys. Lett.* **2016**, *108*, 222101. [CrossRef]
24. Ting, D.Z.; Soibel, A.; Gunapala, S.D. Hole effective masses and subband splitting in type-II superlattice infrared detectors. *Appl. Phys. Lett.* **2016**, *108*, 183504. [CrossRef]
25. Ting, D.Z.; Soibel, A.; Gunapala, S.D. Type-II superlattice hole effective masses. *Infrared Phys. Technol.* **2017**, *84*, 102–106. [CrossRef]
26. Maimon, S.; Wicks, G.W. nBn detector, an infrared detector with reduced dark current and higher operating temperature. *Appl. Phys. Lett.* **2006**, *89*, 151109. [CrossRef]
27. Klipstein, P. Depletion-Less Photodiode with Suppressed Dark Current and Method for Producing the Same. Int. Patent Publication No: WO 2005/004243 A1, 2005.
28. Klipstein, P.C. XBn barrier photodetectors for high sensitivity and high operating temperature infrared sensors. In Proceedings of the Infrared Technology and Applications XXXIV, Orlando, FL, USA, 16–20 March 2008.
29. Ting, D.Z.-Y.; Hill, C.J.; Soibel, A.; Keo, S.A.; Mumolo, J.M.; Nguyen, J.; Gunapala, S.D. A high-performance long wavelength superlattice complementary barrier infrared detector. *Appl. Phys. Lett.* **2009**, *95*, 023508. [CrossRef]
30. Ting, D.Z.-Y.; Soibel, A.; Khoshakhlagh, A.; Nguyen, J.; Höglund, L.; Keo, S.A.; Mumolo, J.M.; Gunapala, S.D. Exclusion, extraction, and junction placement effects in the complementary barrier infrared detector. *Appl. Phys. Lett.* **2013**, *102*, 121109. [CrossRef]
31. Canedy, C.L.; Aifer, E.H.; Vurgaftman, I.; Tischler, J.G.; Meyer, J.R.; Warner, J.H.; Jackson, E.M. Antimonide Type-II "W" Photodiodes with Long-Wave Infrared R0A Comparable to HgCdTe. *J. Electron. Mater.* **2007**, *36*, 852–856. [CrossRef]
32. Nguyen, B.-M.; Hoffman, D.; Delaunay, P.-Y.; Razeghi, M. Dark current suppression in type II InAs/GaSbInAs/GaSb superlattice long wavelength infrared photodiodes with M-structure barrier. *Appl. Phys. Lett.* **2007**, *91*, 163511. [CrossRef]
33. Nguyen, B.M.; Chen, G.; Hoang, A.M.; Abdollahi Pour, S.; Bogdanov, S.; Razeghi, M. Effect of contact doping in superlattice-based minority carrier unipolar detectors. *Appl. Phys. Lett.* **2011**, *99*, 033501. [CrossRef]
34. Wicks, G.W.; Savich, G.R.; Pedrazzani, J.R.; Maimon, S. Infrared detector epitaxial designs for suppression of surface leakage current. In Proceedings of the Quantum Sensing and Nanophotonic Devices VII, San Francisco, CA, USA, 23–28 January 2010.
35. Savich, G.R.; Pedrazzani, J.R.; Sidor, D.E.; Maimon, S.; Wicks, G.W. Dark current filtering in unipolar barrier infrared detectors. *Appl. Phys. Lett.* **2011**, *99*, 121112. [CrossRef]
36. Rodriguez, J.B.; Plis, E.; Bishop, G.; Sharma, Y.D.; Kim, H.; Dawson, L.R.; Krishna, S. nBn structure based on InAs/GaSb type-II strained layer superlattices. *Appl. Phys. Lett.* **2007**, *91*, 043514. [CrossRef]
37. Smith, D.L.; Mailhiot, C. Proposal for strained type II superlattice infrared detectors. *Appl. Phys. Lett.* **1987**, *34*, 663–665. [CrossRef]
38. Osbourn, G.C. Electro-Optical SLS Devices for Operating at New Wavelength Ranges. U.S. Patent No. 4,607,272, 19 August 1986.
39. Osbourn, G.C. InAsSb Strained layer superlattices for long wavelength detector applications. *J. Vac. Sci. Technol. B* **1984**, *2*, 176. [CrossRef]
40. Osbourn, G.C.; Dawson, L.R.; Biefeld, R.M.; Zipperlan, T.E.; Fritz, I.J.; Doyle, B.L. III-V strained layer superlattices for long-wavelength detector applications: Recent progress. *J. Vac. Sci. Technol. A* **1987**, *5*, 3150–3152. [CrossRef]
41. Kurtz, S.R.; Dawson, L.R.; Zipperian, T.E.; Lee, S.R. Demonstration of an InAsSb strained layer superlattice photodiode. *Appl. Phys. Lett.* **1988**, *52*, 1581. [CrossRef]
42. Kurtz, S.R.; Dawson, L.R.; Biefeld, R.M.; Fritz, I.J.; Zipperian, T.E. Long wavelength InAsSb strained layer superlattice photovoltaic infrared detector. *IEEE Elect. Device Lett.* **1989**, *10*, 150–152. [CrossRef]

43. Kurtz, S.R.; Biefeld, R.M.; Zipperlan, T.E. MOCVD-grown InAsSb strained-layer superlattice infrared detectors with photoresponses ≥ 10 μm. *Semicond. Sci. Technol.* **1990**, *5*, S24–S26. [CrossRef]
44. Zhang, Y.-H. Continuous wave operation of InAs/InAsxSb12x midinfrared lasers. *Appl. Phys. Lett.* **1995**, *66*, 118–120. [CrossRef]
45. Tang, P.J.P.; Pullin, M.J.; Chung, S.J.; Phillips, C.C.; Stradling, R.A.; Norman, A.G.; Li, Y.B.; Hart, L. 4–11 pm infrared emission and 300 K light emitting diodes from arsenic-rich InAs/InAs1-xSbx strained layer superlattices. *Semicond. Sci. Technol.* **1995**, *10*, 1177–1180. [CrossRef]
46. Ciesla, C.M.; Murdin, B.N.; Pidgeon, C.R.; Stradling, R.A.; Phillips, C.C.; Livingstone, M.; Galbraith, I.; Jaroszynski, D.A.; Langerak, C.J.G.M.; Tang, P.J.P.; et al. Suppression of Auger recombination in arsenic-rich InAs1-xSbx strained layer superlattices. *J. Appl. Phys.* **1996**, *80*, 2994. [CrossRef]
47. Pullin, M.J.; Hardaway, H.R.; Heber, J.D.; Phillips, C.C.; Yuen, W.T.; Stradling, R.A. Room-temperature InAsSb strained-layer superlattice light-emitting diodes at λ = 4.2 μm with AlSb barriers for improved carrier confinement. *Appl. Phys. Lett.* **1999**, *74*, 2384–2386.
48. Chow, D.H.; Miles, R.H.; Söderström, J.R.; McGill, T.C. Growth and characterization of InAs/Ga1-xInxSb strained-layer superlattices. *Appl. Phys. Lett.* **1990**, *56*, 1418–1420. [CrossRef]
49. Lackner, D.; Pitts, O.J.; Steger, M.; Yang, A.; Thewalt, M.L.W.; Watkins, S.P. Strain balanced InAs/InAsSb superlattice structures with optical emission to 10 μm. *Appl. Phys. Lett.* **2009**, *95*, 081906. [CrossRef]
50. Haddadi, A.; Dehzangi, A.; Adhikary, S.; Chevallier, R.; Razeghi, M. Background–limited long wavelength infrared InAs/InAs1− xSbx type-II superlattice-based photodetectors operating at 110 K. *APL Mater.* **2017**, *5*, 035502. [CrossRef]
51. Chevallier, R.; Haddadi, A.; Razeghi, M. Dark current reduction in microjunction-based double electron barrier type-II InAs/InAsSb superlattice long-wavelength infrared photodetectors. *Sci. Rep.* **2017**, *7*, 12617. [CrossRef] [PubMed]
52. Hoang, A.M.; Chen, G.; Chevallier, R.; Haddadi, A.; Razeghi, M. High performance photodiodes based on InAs/InAsSb type-II superlattices for very long wavelength infrared detection Appl. *Phys. Lett.* **2014**, *104*, 251105.
53. Haddadi, A.; Chevallier, R.; Chen, G.; Hoang, A.M.; Razeghi, M. Bias-selectable dual-band mid-/long-wavelength infrared photodetectors based on InAs/InAs1−xSbx type-II superlattices. *Appl. Phys. Lett.* **2015**, *106*, 011104. [CrossRef]
54. Ting, D.Z.; Soibel, A.; Khoshakhlagh, A.; Keo, S.A.; Rafol, S.B.; Höglund, L.; Luong, E.M.; Fisher, A.M.; Hill, C.J.; Gunapala, S.D. Development of InAs/InAsSb Type II Strained-Layer Superlattice Unipolar Barrier Infrared Detectors. *J. Electron. Mater.* **2019**, *48*, 6145–6151. [CrossRef]
55. Ting, D.Z.; Soibel, A.; Khoshakhlagh, A.; Keo, S.A.; Rafol, S.B.; Fisher, A.M.; Hill, C.J.; Luong, E.M.; Pepper, B.J.; Gunapala, S.D. The emergence of InAs/InAsSb type-II strained layer superlattice barrier infrared detectors. In Proceedings of the SPIE 11002, Infrared Technology and Applications XLV, Baltimore, MD, USA, 14–18 April 2019; p. 110020F.
56. Ting, D.Z.; Soibel, A.; Khoshakhlagh, A.; Rafol, S.B.; Keo, S.; Höglund, L.; Fisher, A.M.; Luong, E.M.; Gunapala, S.D. Mid-wavelength high operating temperature barrier infrared detector and focal plane array. *Appl. Phys. Lett.* **2018**, *113*, 021101. [CrossRef]
57. Ting, D.Z.; Rafol, S.B.; Keo, S.; Nguyen, J.; Khoshakhlagh, A.; Soibel, A.; Hoglund, L.; Fisher, A.M.; Luong, E.M.; Mumolo, J.; et al. InAs/InAsSb Type-II Superlattice Mid-Wavelength Infrared Focal Plane Array with Significantly Higher Operating Temperature Than InSb. *IEEE Photon. J.* **2018**, *10*, 6804106. [CrossRef]
58. Tennant, W.E. "Rule 07" Revisited: Still a Good Heuristic Predictor of p/n HgCdTe Photodiode Performance? *J. Electron. Mater.* **2010**, *39*, 1030. [CrossRef]
59. Klipstein, P.C.; Aronov, D.; Ben Ezra, M.; Barkai, I.; Berkowicz, E.; Brumer, M.; Fraenkel, R.; Glozman, A.; Grossman, S.; Jacobsohn, E.; et al. Recent progress in InSb based quantum detectors in Israel. *Infrared Phys. Technol.* **2013**, *59*, 172–181. [CrossRef]
60. Olson, B.V.; Shaner, E.A.; Kim, J.K.; Klem, J.F.; Hawkins, S.D.; Murray, L.M.; Prineas, J.P.; Flatté, M.E.; Boggess, T.F. Time-resolved optical measurements of minority carrier recombination in a mid-wave infrared InAsSb alloy and InAs/InAsSb superlattice. *Appl. Phys. Lett.* **2012**, *101*, 092109. [CrossRef]
61. Höglund, L.; Ting, D.Z.; Khoshakhlagh, A.; Soibel, A.; Hill, C.J.; Fisher, A.; Keo, S.; Gunapala, S.D. Influence of radiative and non-radiative recombination on the minority carrier lifetime in midwave infrared InAs/InAsSb superlattices. *Appl. Phys. Lett.* **2013**, *103*, 221908. [CrossRef]

62. Donetsky, D.; Svensson, S.P.; Vorobjev, L.E.; Belenky, G. Carrier lifetime measurements in short-period InAs/GaSb strained-layer superlattice structures. *Appl. Phys. Lett.* **2009**, *95*, 212104. [CrossRef]
63. Prins, A.D.; Lewis, M.K.; Bushell, Z.L.; Sweeney, S.J.; Liu, S.; Zhang, Y.-H. Evidence for a defect level above the conduction band edge of InAs/InAsSb type-II superlattices for applications in efficient infrared photodetectors. *Appl. Phys. Lett.* **2015**, *106*, 171111. [CrossRef]
64. Grein, C.H.; Flatté, M.E.; Ehrenreich, H. Comparison of Ideal InAs/InAs1-xSbx and InAs/InxGa1-xSb Superlattice IR Detectors. In Proceedings of the Third International Symposium on Long-Wavelength Infrared Detectors and Arrays: Physics and Applications, Chicago, IL, USA, 8–13 October 1995; Li, S.S., Liu, H.C., Beck, W.A., Tidrow, M.Z., Singh, A., Eds.; Electrochemical Society: Pennington, NJ, USA, 1995; Volume 95–28, pp. 211–218.
65. Sidor, D.E.; Savich, G.R.; Wicks, G.W. Surface Leakage Mechanisms in III–V Infrared Barrier Detectors. *J. Electron. Mater.* **2016**, *45*, 4663–4667. [CrossRef]
66. Du, X.; Marozas, B.T.; Savich, G.R.; Wicks, G.W. Defect-related surface currents in InAs-based nBn infrared detectors. *J. Appl. Phys.* **2018**, *123*, 214504. [CrossRef]
67. Marozas, B.T.; Hughes, W.D.; Du, X.; Sidor, D.E.; Savich, G.R.; Wicks, G.W. Surface dark current mechanisms in III-V infrared photodetectors. *Opt. Mater. Express* **2018**, *8*, 1419–1424. [CrossRef]
68. Haddadi, A.; Dehzangi, A.; Chevallier, R.; Adhikary, S.; Razeghi, M. Bias-selectable nBn dual-band long-/very long-wavelength infrared photodetectors based on InAs/InAs1-xSbx/AlAs1-xSbx type-II superlattices. *Sci. Rep.* **2017**, *7*, 3379. [CrossRef] [PubMed]
69. Lubyshev, D.; Fastenau, J.M.; Kattner, M.; Frey, P.; Nelson, S.A.; Liu, A.W.K.; Martinez, B.; Furlong, M.J. T2SL Mid- and long- wave infrared photodetector structures grown on (211)A, (211)B, and (311)A GaSb substrates. In Proceedings of the SPIE 11002, Infrared Technology and Applications XLV, Baltimore, MD, USA, 14–18 April 2019; pp. 11002–11023.
70. Ting, D.Z.; Khoshakhlagh, A.; Soibel, A.; Gunapala, S.D. Long Wavelength InAs/InAsSb Infrared Superlattice Challenges: A Theoretical Investigation. *J. Electron. Mater.* 2020. [CrossRef]
71. Ciani, A.J.; Grein, C.H.; Irick, B.; Miao, M.S.; Kioussis, N. Molecular dynamics growth modeling of InAs1−xSbx-based type-II superlattice. *Opt. Eng.* **2017**, *56*, 091609. [CrossRef]
72. Haugan, H.J.; Brown, G.J.; Peoples, J.A. On the study of antimony incorporation in InAs/InAsSb superlattices for infrared sensing. *J. Vac. Sci. Technol. B* **2017**, *35*, 02B107. [CrossRef]
73. Sarney, W.L.; Svensson, S.P.; Yakes, M.K.; Xu, Y.; Donetsky, D.; Belenky, G. Ultra-short period Ga-free superlattice growth on GaSb. *J. Appl. Phys.* **2018**, *124*, 035304. [CrossRef]
74. Kanedy, K.; Lopez, F.; Wood, M.R.; Gmachl, C.F.; Weimer, M.; Klem, J.F.; Hawkins, S.D.; Shaner, E.A.; Kim, J.K. Visualizing period fluctuations in strained-layer superlattices with scanning tunneling microscopy. *Appl. Phys. Lett.* **2018**, *112*, 042105. [CrossRef]
75. Esaki, L.; Chang, L.L.; Mendez, E.E. Polytype Superlattices and Multi-Heterojunctions. *Jpn. J. Appl. Phys.* **1981**, *20*, L529–L532. [CrossRef]
76. Delaunay, P.Y.; Nguyen, B.M.; Hoffman, D.; Huang, E.K.W.; Razeghi, M. Background Limited Performance of Long Wavelength Infrared Focal Plane Arrays Fabricated From M-Structure InAs–GaSb Superlattices. *IEEE J. Quantum Electron.* **2009**, *45*, 157–162. [CrossRef]
77. Salihoglu, O.; Muti, A.; Kutluer, K.; Tansel, T.; Turan, R.; Ergun, Y.; Aydinli, A. "N" structure for type-II superlattice photodetectors. *Appl. Phys. Lett.* **2012**, *101*, 073505. [CrossRef]
78. Hao, R.; Ren, Y.; Liu, S.; Guo, J.; Wang, G.; Xu, Y.; Niu, Z. Fabrication and characterization of high lattice matched InAs/InAsSb superlattice infrared photodetector. *J. Cryst. Growth* **2017**, *470*, 33. [CrossRef]
79. Craig, A.P.; Al-Saymari, F.; Jain, M.; Bainbridge, A.; Savich, G.R.; Golding, T.; Krier, A.; Wicks, G.W.; Marshall, A.R. Resonant cavity enhanced photodiodes on GaSb for the mid-wave infrared. *Appl. Phys. Lett.* **2019**, *114*, 151107. [CrossRef]
80. Durlin, Q.; Perez, J.P.; Cerutti, L.; Rodriguez, J.B.; Cerba, T.; Baron, T.; Tournie, E.; Christol, P. Midwave infrared barrier detector based on Ga-free InAs/InAsSb type-II superlattice grown by molecular beam epitaxy on Si substrate. *Infrared Phys. Technol.* **2019**, *96*, 39–43. [CrossRef]
81. Ariyawansa, G.; Duran, J.; Reyner, C.; Scheihing, J. InAs/InAsSb Strained-Layer Superlattice Mid-Wavelength Infrared Detector for High-Temperature Operation. *Micromachines* **2019**, *10*, 806. [CrossRef]

82. Deng, G.; Song, X.; Pan, M.; Xiao, T.; Luo, Z.; Chen, N.; Yang, W.; Zhang, Y. Upside-down InAs/InAs1-xSbx type-II superlattice-based nBn mid-infrared photodetectors with an AlGaAsSb quaternary alloy barrier. *Opt. Express* **2020**, *28*, 13616–13624. [CrossRef]
83. Manyk, T.; Michalczewski, K.; Murawski, K.; Martyniuk, P.; Rutkowski, J. InAs/InAsSb Strain-Balanced Superlattices for Longwave Infrared Detectors. *Sensors* **2019**, *19*, 1907. [CrossRef]
84. Michalczewski, K.; Kubiszyn, Ł.; Martyniuk, P.; Wu, C.H.; Jureńczyk, J.; Grodecki, K.; Benyahia, D.; Rogalski, A.; Piotrowski, J. Demonstration of HOT LWIR T2SLs InAs/InAsSb photodetectors grown on GaAs substrate. *Infrared Phys. Technol.* **2018**, *95*, 222. [CrossRef]
85. Michalczewski, K.; Martyniuk, P.; Kubiszyn, L.; Wu, C.H.; Wu, Y.R.; Jurenczyk, J.; Rogalski, A.; Piotrowski, J. Demonstration of the Very Long Wavelength Infrared Type-II Superlattice InAs/InAsSb GaAs Immersed Photodetector Operating at Thermoelectric Cooling. *IEEE Electron Device Lett.* **2019**, *40*, 1396–1398. [CrossRef]
86. Letka, V.; Bainbridge, A.; Craig, A.P.; Al-Saymari, F.; Marshall, A.R.J. Resonant cavity-enhanced photodetector incorporating a type-II superlattice to extend MWIR sensitivity. *Opt. Express* **2019**, *27*, 23970–23980. [CrossRef] [PubMed]
87. Yeche, A.; Boulard, F.; Cervera, C.; Perez, J.P.; Rodriguez, J.B.; Christol, P.; Gravrand, O. Development of Electron Beam Induced Current for diffusion length determination of VLWIR HgCdTe and MWIR T2SL based photodetectors. *Infrared Phys. Technol.* **2018**, *95*, 170–176. [CrossRef]
88. Rhiger, D.R.; Smith, E.P. Carrier Transport in the Valence Band of nBn III-V Superlattice Infrared Detectors. *J. Electron. Mater.* **2019**, *48*, 6053–6062. [CrossRef]
89. Casias, L.K.; Morath, C.P.; Steenbergen, E.H.; Umana-Membreno, G.A.; Webster, P.T.; Logan, J.V.; Kim, J.K.; Balakrishnan, G.; Faraone, L.; Krishna, S. Vertical carrier transport in strain-balanced InAs/InAsSb type-II superlattice material. *Appl. Phys. Lett.* **2019**, *116*, 182109. [CrossRef]
90. Tsai, C.Y.; Zhang, Y.; Ju, Z.; Zhang, Y.H. Study of vertical hole transport in InAs/InAsSb type-II superlattices by steady-state and time-resolved photoluminescence spectroscopy. *Appl. Phys. Lett.* **2020**, *116*, 201108. [CrossRef]
91. Huang, Y.; Ryou, J.-H.; Dupuis, R.D.; D'Costa, V.R.; Steenbergen, E.H.; Fan, J.; Zhang, Y.-H.; Petschke, A.; Mandl, M.; Chuang, S.-L. Epitaxial growth and characterization of InAs/GaSb and InAs/InAsSb type-II superlattices on GaSb substrates by metalorganic chemical vapor deposition for long wavelength infrared photodetectors. *J. Cryst. Growth* **2011**, *314*, 92–96. [CrossRef]
92. Wu, D.H.; Durlin, Q.; Dehzangi, A.; Zhang, Y.; Razeghi, M. High quantum efficiency mid-wavelength infrared type-II InAs/InAs1−xSbx superlattice photodiodes grown by metal-organic chemical vapor deposition. *Appl. Phys. Lett.* **2019**, *114*, 011104. [CrossRef]
93. Ning, Z.-D.; Liu, S.-M.; Luo, S.; Ren, F.; Wang, F.; Yang, T.; Liu, F.-Q.; Wang, Z.G.; Zhao, L.-C. Growth and characterization of InAs/InAsSb superlattices by metal organic chemical vapor deposition for mid-wavelength infrared photodetectors. *Mater. Lett.* **2016**, *164*, 213. [CrossRef]
94. Wu, D.H.; Dehzangi, A.; Zhang, Y.Y.; Razeghi, M. Demonstration of long wavelength infrared type-II InAs/InAs1−xSbx superlattices photodiodes on GaSb substrate grown by metalorganic chemical vapor deposition. *Appl. Phys. Lett.* **2018**, *112*, 241103. [CrossRef]
95. Delli, E.; Letka, V.; Hodgson, P.D.; Repiso, E.; Hayton, J.P.; Craig, A.P.; Lu, Q.; Beanland, R.; Krier, A.; Marshall, A.R.J.; et al. Mid-Infrared InAs/InAsSb Superlattice nBn Photodetector Monolithically Integrated onto Silicon. *ACS Photon.* **2019**, *6*, 538. [CrossRef]
96. Fastenau, J.M.; Lubyshev, D.; Nelson, S.A.; Fetters, M.; Krysiak, H.; Zeng, J.; Kattner, M.; Frey, P.; Liu, A.W.K.; Morgan, A.O.; et al. Direct MBE Growth of Metamorphic nBn Infrared Photodetectors on 150 mm Ge-Si Substrates for Heterogeneous Integrations. *J. Vac. Sci. Technol. B* **2019**, *37*, 031216. [CrossRef]
97. Baril, N.F.; Bandara, S.; Zuo, D.; Brown, A.; Almeida, L.A.; Tidrow, M. Growth of III-V Infrared Detector Device Layers on the AlSb Lattice via Metamorphic Buffer Layers. Paper 11002-19, SPIE DCS 2019. Available online: https://spie.org/SI19/conferencedetails/infrared-technology-applications#session-5 (accessed on 26 October 2020).
98. Lubyshev, D.; Fastenau, J.M.; Kattner, M.; Frey, P.; Nelson, S.A.; Flick, R.; Rogers, M.; Liu, A.W.K.; Flint, P.; Faleev, N. Effect of substrate orientation on Sb-based MWIR photodetector characteristics. *Infrared Phys. Technol.* **2018**, *95*, 27–32. [CrossRef]

99. Soibel, A.; Ting, D.Z.; Rafol, S.B.; Fisher, A.M.; Keo, S.A.; Khoshakhlagh, A.; Gunapala, S.D. Mid-wavelength infrared InAsSb/InAs nBn detectors and FPAs with very low dark current density. *Appl. Phys. Lett.* **2019**, *114*, 161103. [CrossRef]
100. Pagano, T.S.; Rider, D.; Rud, M.; Ting, D.; Yee, K. Measurement approach and design of the CubeSat Infrared Atmospheric Sounder (CIRAS). In Proceedings of the SPIE 9978, CubeSats and NanoSats for Remote Sensing, San Diego, CA, USA, 28 August–1 September 2016; p. 997806.
101. Tidrow, M.Z.; Reago, D.A., Jr. VISTA video and overview (Conference Presentation). In Proceedings of the SPIE 10177, Infrared Technology and Applications XLIII, Anaheim, CA, USA, 9–13 April 2017; p. 101770M.
102. Tidrow, M.; Bandara, S.; Aitcheson, L.; Zheng, L.; Baril, N.; Williams, A. III-V infrared focal plane array development in US (Conference Presentation). In Proceedings of the SPIE 10624, Infrared Technology and Applications XLIV, Orlando, FL, USA, 15–19 April 2018; p. 106240P.
103. Flint, P.; Dallas, G.; Bollaert, A. Production manufacturing of 5″ diameter gallium antimonide substrates (Conference Presentation). In Proceedings of the SPIE 10177, Infrared Technology and Applications XLIII, Anaheim, CA, USA, 9–13 April 2017; p. 101770P.
104. Pinsukanjana, P.R.; Andresen, B.F. Large diameter Epi-ready GaSb substrate manufacturing (Conference Presentation). In Proceedings of the SPIE 10177, Infrared Technology and Applications XLIII, Anaheim, CA, USA, 9–13 April 2017; p. 101770Q.
105. Liu, A.W.K.; Lubyshev, D.; Qiu, Y.; Fastenau, J.M.; Wu, Y.; Furlong, M.J.; Tybjerg, M.J.; Martinez, R.J.; Mowbray, A.; Smith, B. MBE growth of Sb-based bulk nBn infrared photodetector structures on 6-inch GaSb substrates. In Proceedings of the SPIE 9451, Infrared Technology and Applications XLI, Baltimore, MD, USA, 20–24 April 2015; p. 94510T.
106. Liu, A.W.K.; Lubyshev, D.; Fastenau, J.M.; Nelson, S.; Kattner, M.; Frey, P. Molecular beam epitaxial growth and characterization of large-format GaSb-based IR photodetector structures. *Opt. Mater. Express* **2018**, *8*, 1282. [CrossRef]
107. Nelson, S.A.; Fastenau, J.M.; Lubyshev, D.; Kattner, M.; Frey, P.; Liu, A.W.K.; Martinez, B.; Furlong, M.J. Large format multi-wafer production of LWIR photodetector structures on 150 mm GaSb substrates by MBE. In Proceedings of the SPIE 11407, Infrared Technology and Applications XLVI, Online Only, CA, USA, 27 April–8 May 2020; p. 114070F.
108. Pinsukanjana, P.R.; Andresen, B.F. Multi-wafer production MBE capabilities for Sb-based type-II SLS IR detectors (Conference Presentation). In Proceedings of the SPIE 10177, Infrared Technology and Applications XLIII, Anaheim, CA, USA, 9–13 April 2017; p. 101770R.

Publisher's Note: MDPI stays neutral with regard to jurisdictional claims in published maps and institutional affiliations.

© 2020 by the authors. Licensee MDPI, Basel, Switzerland. This article is an open access article distributed under the terms and conditions of the Creative Commons Attribution (CC BY) license (http://creativecommons.org/licenses/by/4.0/).

micromachines

MDPI

Review

Overview of Ultrasound Detection Technologies for Photoacoustic Imaging

Rayyan Manwar [1,2], Karl Kratkiewicz [2] and Kamran Avanaki [1,2,3,*]

1 Richard and Loan Hill Department of Bioengineering, University of Illinois at Chicago, Chicago, IL 60607, USA; r.manwar@wayne.edu
2 Department of Biomedical Engineering, Wayne State University, Detroit, MI 48201, USA; karl.kratkiewicz@wayne.edu
3 Department of Dermatology, University of Illinois at Chicago, Chicago, IL 60607, USA
* Correspondence: mrn.avanaki@wayne.edu; Tel.: +1-313-577-0703

Received: 12 June 2020; Accepted: 14 July 2020; Published: 17 July 2020

Abstract: Ultrasound detection is one of the major components of photoacoustic imaging systems. Advancement in ultrasound transducer technology has a significant impact on the translation of photoacoustic imaging to the clinic. Here, we present an overview on various ultrasound transducer technologies including conventional piezoelectric and micromachined transducers, as well as optical ultrasound detection technology. We explain the core components of each technology, their working principle, and describe their manufacturing process. We then quantitatively compare their performance when they are used in the receive mode of a photoacoustic imaging system.

Keywords: ultrasound transducer; photoacoustic imaging; piezoelectric; micromachined; CMUT; PMUT; optical ultrasound detection

1. Introduction

Ultrasound transducers are devices that convert ultrasound pressure waves into electrical signal. In an ultrasound imaging machine the transducer is a transceiver device: the waves propagated from an ultrasound transducer are backscattered/reflected from an impedance mismatch in the tissue and received by the same transducer; the strength of the received pressure waves is in the range of 0.1~4 MPa [1]. Another modality that directly benefits from ultrasound transducer technology is photoacoustic imaging (PAI). PAI is an emerging modality that uses a combination of optical excitation and acoustic detection for visualizing vascular, functional, and molecular changes within living tissue [2–11]. As opposed to the optical imaging modalities such as optical coherence tomography [12] that employs ballistic photons, PAI uses diffused photons providing significantly deeper penetration. In PAI, thermoelastic expansion of tissue chromophores occurs when irradiated by a nanosecond pulsed laser—resulting in emission of acoustic waves that are then detected by ultrasound transducers for image formation [9,13–17]; in PAI the ultrasound transducer is a receiver device. The strength of the acoustic waves generated from the chromophores is around 800 PA·mK^{-1} [4]. The strength of the generated pressure in PAI depends on the absorption coefficient of the chromophores, the light fluence, and the characteristics of the ultrasound transducer. The lower range of the generated pressure waves in PAI compared to ultrasound imaging signifies the importance of an efficient and effective ultrasound detection technology [18]. In PAI, where the optically induced ultrasound pressure is typically weak [17], the primary requirement of the detection unit is to have a high sensitivity and a large acceptance angle over a wide range of spectral bandwidth.

Acoustic and optical detection methods are complementary technologies that together have solved many unmet industrial and clinical needs. The contactless nature and the wavelength selectivity capability to study a particular target in the tissue (e.g., enabling functional sensing) are advantages of optical sensing over acoustic sensing, and the less penetration depth of optical sensing compared to acoustic detection technology is its disadvantage; although utilizing short wavelengths (such as X-ray), deep sensing applications are possible at a cost of ionizing the imaging target. In photoacoustic sensing, the advantages of both technologies are utilized: wavelength selectivity and adequate penetration depth; to address different unmet needs, photoacoustic technology has been implemented for different wavelengths from X-ray to infrared (IR). Due to the widely used intermediate penetration depth achieved by IR light, higher sensitivity of IR devices, and the fast growing advancement of IR optical components, most of photoacoustic systems are implemented in this regime. Infrared photoacoustic technology has been successfully used in both preclinical (to study small animal brain [19–22], eye [23–26], and skin [27–29]) and clinical (to detect breast cancer [30–32], cervical cancer [33,34], skin melanoma [35,36], and brain tumor [37,38]) applications.

Based on the ultrasound detection mechanism, transducers can be categorized into two main categories: physical ultrasound transducers and optical ultrasound detection. The physical transducers can further be classified as: (i) conventional piezoelectric, and (ii) micromachined (capacitive or piezoelectric).

Several review studies have been conducted on ultrasound transducer technologies [39–46]. However, these studies were primarily focused on ultrasound imaging and specific to one or two technologies. The purpose of this review is to study the effectiveness of various ultrasound transducer technologies, recognize their pros and cons, and learn about their performance when they are used in the receive mode of a photoacoustic imaging system. This review does not cover the ultrasound transducer technologies that are specifically used for industrial or intravascular applications.

The search protocol used for this review study is as follows. A PubMed database search of "transducer" AND "ultrasound" AND "photoacoustic" AND "imaging" yielded 216 results with 196 published in the last ten years. We have narrowed down the search by "piezoelectric" (45 results) and "micromachined" (29 results). Capacitive micromachined ultrasonic transducers (CMUTs) and piezoelectric micromachined ultrasound transducers (PMUTs) have been utilized in 22 and 7 articles, respectively. Moreover, 4 articles were found relevant to photoacoustic imaging among articles on optical ultrasound detection. In this study, we have reviewed a total of 189 articles.

The organization of the manuscript is as follows. First, the general design characteristics of physical ultrasound transducers are discussed. We then investigate the physical ultrasound transducers along with a quantitative analysis of their performance. Next, we discuss optical ultrasound detection technologies and present their corresponding specifications. Finally, we summarize the pros and cons of various ultrasound detection technologies and discuss their performance in photoacoustic imaging applications.

2. Ultrasound Transducer Characteristics

The design parameters in an ultrasound transducer are classified into two categories: (i) geometric characteristics of layers (width, length, thickness, and specific to arrays including the number of elements, kerf, and pitch size [47]), (ii) material properties (such as coupling coefficient, elastic modulus, Poisson's ratio, density, stress coefficient, stiffness constant, acoustic impedance, and dielectric constant) used for each section. By adjusting these parameters, a transducer with a desired sensitivity, center frequency, and bandwidth is obtained. If cost is a deciding factor, sensitivity and bandwidth of the transducer may be affected. The manufacturing cost of a transducer largely depends on the fabrication process and the number of attempts needed to obtain required specifications [9].

Electromechanical coupling coefficient of the material represents the coupling efficiency of the transducer. In receive mode, this coefficient can be defined as the ratio between the electrical energy induced and the mechanical energy applied to the sensing material [48]. Coupling coefficient is primarily determined based on the inherent material properties of the transducer sensing elements such as stress coefficient, stiffness constant, acoustic impedance, and dielectric constant. Stress coefficient and stiffness constant are two mechanical properties that determine the elasticity of the sensing material.

A highly elastic material with thermal stability is desired to build a transducer with a wide frequency range and low mechanical loss. Acoustic impedance determines the compliance of the transducer material to the target tissue material. Acoustic waves can be transmitted efficiently through the propagating medium when there is less acoustic impedance mismatch between the transducer material and the imaging target medium. Reduced acoustic impedance mismatch improves the signal-to-noise ratio (SNR) of the signal converted from the received pressure waves. In addition to acoustic impedance, there is the electrical impedance match between the transducer material and the back-end electronics (i.e., signal routing, data acquisition unit, amplifiers). Electrical impedance also affects the power transmission efficiency and SNR of the converted signal. Such electrical match is achieved using a matching network that can be realized using a material with high dielectric constant [49,50]. Moreover, a high dielectric constant is essential to improve the coupling coefficient [51].

The receive sensitivity of an ultrasound transducer is typically represented as a ratio of the detected electrical signal amplitude (in the range of micro- to milli-volts) and applied acoustic pressure (in the range of pascal to kilopascal). In photoacoustic imaging, the sensitivity is represented by noise equivalent pressure (NEP) [52], a frequency dependant metric with a unit of $Pa{\cdot}Hz^{-1/2}$. NEP is defined as the photoacoustic pressure at the imaging target that generates a transducer output equal to the noise amplitude [53]. We used sensitivity (mV/kPa) for quantitative comparison between the physical transducers and NEP ($Pa{\cdot}Hz^{-1/2}$) for transducers that work based on optical ultrasound detection methods.

An ideal ultrasound detection device should possess the following attributes: sufficiently high electromechanical coupling coefficient, an acoustic impedance that is close to tissue impedance, a large dielectric constant, low electrical and mechanical losses, low stiffness, and high thermal stability, that all together leads to a transducer with a high sensitivity over a wide spectral bandwidth.

3. Physical Ultrasound Transducer Technologies

3.1. Piezoelectric Transducers

Piezoelectric ultrasound transducers are the most widely manufactured and clinically available transducers that are integrated in commercial ultrasound systems [39,54,55]. The main component of a piezoelectric ultrasound transducer is piezo-material that operates based on converse and direct piezoelectric effect. In transmission mode of an ultrasound transducer, the generated acoustic waves are a result of the transient expansion and contraction of a piezo-material when exposed to an alternating electric field across the piezo-electrodes [56]. In receive mode, the incident acoustic pressure waves deform the piezo-material, and are measured in terms of the potential difference across the piezo-electrodes induced by the deformation [57,58]. A cross section of a piezoelectric linear array transducer is shown in Figure 1a. Piezoelectric ultrasound transducer elements are usually manufactured with a matching layer to reduce the impedance mismatch between the imaging target and backing layer to suppress the back scattered ringing effect. Among piezo-materials, naturally occurring crystals (quartz [59]), are seldom used in manufacturing transducers because of their weak piezoelectric performance, low dielectric and elastic properties, and low stability [54]. Engineered single crystals (such as lead magnesium niobate–lead titanate (PMN–PT) [60] and lead zinc niobate–lead titanate (PZN–PT) [61]) exhibit a high coupling coefficient and a large bandwidth that can specifically be valuable to photoacoustic imaging applications, however, the manufacturing process of these transducers is complex, expensive, and time consuming.

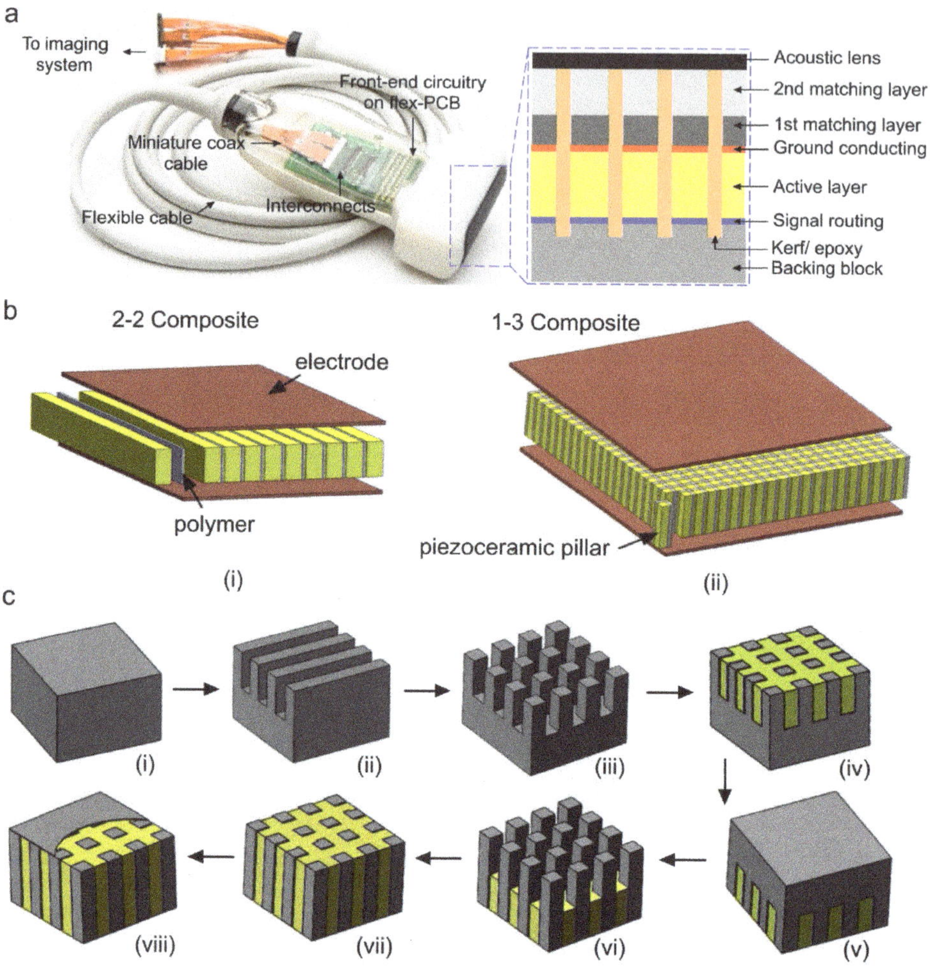

Figure 1. Geometric characteristics and manufacturing steps of a piezoelectric linear array ultrasound transducer. (**a**) A photograph of a piezoelectric ultrasound imaging probe; the cross-section of the sensing layer is provided in the blue dotted box [39], (**b**) structural difference between 2-2, 1-3 composite material when used in an ultrasound array transducer [62], (**c**) process flow of conventional dice and fill (DF) fabrication method using 1-3 composite and epoxy filling that includes: (i) piezoceramic material, (ii) dice in x direction, (iii) dice in y direction, (iv) epoxy filling, (v) reverse, (vi) backside dicing in x and y directions, (vii) 2nd epoxy filling, and (viii) deposit conductive layer. Reprinted with permission from [63].

The most popular piezoelectric materials are piezoceramics (such as barium titanate ($BaTiO_3$) [64], lithium niobite ($LiNbO_3$) [65,66], lead zirconate titanate (PZT) [67], zinc oxide (ZnO) [68]), and polymers (such as polyvinylidene difluoride (PVDF) [69]). Piezoceramics consist of randomly oriented crystallites separated by grain boundaries. They offer strong piezoelectric properties along the polarization axes, and are less expensive than polymers. Polymers such as PVDF as piezo-material alongside their copolymer trifluoroethylene (TrFE) as the thin electrode have also been found to be effective for producing high frequency transducers due to their low stiffness and improved adhesion (compliance) when compared to traditional sputtered thick metal electrodes [70–74]. With these polymers, low acoustic impedance (i.e., close to the tissue impedance) can be achieved at the cost of low energy conversion. To further improve the quality of piezoelectric transducers, composite materials have been developed [75]. The piezoelectric composite consists of a piezoelectric phase (piezo-ceramic) and a polymer phase (epoxy resin), with a certain connection mode, a certain volume or mass ratio, and a certain spatial geometric distribution [76]. Among different materials, PZT-epoxy resin-based composites have been the dominant material to realize the active elements of transducers in piezoelectric transducers [77–80]. Piezo-composites are classified according to respective phase connectivity (0, 1, 2, or 3) through which the phase is continuous. Since, there are two phases in piezo-composites they are referred by 2-digit numbers [76]. The first digit references the piezoelectric phase and the second digit references the polymer phase. Out of 10 conventional different combinations of connectivity [81,82], piezoelectric 1–3 [9,13] and 2–2 [14] composites are commonly used in transducer technology and are proven to exhibit high coupling coefficient with low-acoustic impedance and low stiffness, leading to improved sensitivity compared to monolithic piezo-materials [62,83]. Structural schematics of 1-3 and 2-2 piezo-composites are shown in Figure 1b. The composites are limited to low energy and low temperature applications due to their inherently low mechanical quality factor and thermal conductivity. A quantitative comparison among different types of piezo-material in terms of their determinant properties is provided in Table 1. Speed of sound (SOS) in biological tissues are in the range of 1450–1580 ms^{-1}, thus it is desirable to choose a piezo-material with similar SOS. Table 1 shows that piezo-composite materials provide acoustic impedance and SOS similar to those of biological tissues, with a higher coupling coefficient as compared to polymer or ceramic based piezo-materials; that justifies the use of composites as the preferred piezo layer in ultrasound transducers.

Table 1. Material properties of widely used piezoelectric materials in manufacturing of ultrasound transducers [65,66,76,84–86].

Piezo-materials	Acoustic Impedance (MRayl)	Coupling Coefficient	Relative Permittivity	Density ($kg{\cdot}m^{-3}$)	Speed of Sound ($m{\cdot}s^{-1}$)
Quartz	13.3	0.093	4.5	2648	5000
$LiNbO_3$	39	0.49	39	4700	7360
PZT	33.7	0.51	1470–1700	7500	4580
PMN-PT	37.1	0.58	680–800	8060	4610
PVDF	3.9	0.12–0.29	5–13	1780	2200
1-3 Composite	9	0.6	450	3673	1540

Piezoelectric transducers can be developed as single elements or aggregated into an array (e.g., linear, convex, arc, ring, and spherical). A list of ultrasound transducer arrays that have been used in different photoacoustic imaging applications is given in Table 2. Conventionally, the arrays are realized through dice and fill method (DF) [75]. The DF method involves making a series of parallel cuts on a piece of bulk piezoelectric material with a mechanical dicing saw (Figure 1c shows the steps of a conventional DF fabrication method using 1-3 composite and epoxy filling). The material is then diced in the perpendicular direction to produce posts with a rectangular cross section. The diced material is backfilled with a polymer, then the base ceramic support is removed by lapping polishing [62,63,76]. For 2-2 composite, step 3 is skipped and the remaining steps are similar to those for 1-3 composite. Other alternative methods to make piezoelectric transducers include the interdigital bonding technique, stacked plates or lamination techniques, fiber processing, and laser machining [75].

Table 2. Different configurations of piezoelectric ultrasound transducer arrays that are used in clinical applications of photoacoustic imaging. BW: bandwidth.

Application	Element no.	Configuration	Center Frequency (MHz)	BW (%)	Ref
Breast cancer	588	Hemispherical	1	130	[87]
	512	Hemispherical	2	>100	[88]
	64	Arc	1.5	130	[89]
Dermatology	Single	Spherically focused	54.2	97	[90]
			102.8	105	[91]
Vascular	Single	Focused	50	70	[92]
	256	Linear	21	66	[93]
Carotid vessel	128	Linear	5	80	[94]
	Single	Spherically focused	100	80	[95]
Musculoskeletal	32	Unfocused	6.25	80	[96]
	128	Linear	11.25	75	[97]
Adipose tissue	256	Curved	5	60	[98,99]
Thyroid	192	Linear	5.8	82.7	[100]
	64	Arc	7.5	NA	[101]
Gynecology &Urology	128	Microconvex	6.5	NA	[102,103]

NA: not available.

One design constraint in piezoelectric transducer arrays is that the center frequency is inversely proportional to the thickness. Simultaneously, the element length (l) and width (w) to thickness (t) ratio of l:$t \geq 10$ and w:$t \leq 0.5$ must be maintained [104–106]. Despite the simplicity of DF method, a maximum kerf width of 10 to 15 μm can be achieved using this method and hence, manufacturing high frequency transducers (center frequency: >20 MHz) is difficult [107]. Other alternative methods such as interdigital bonding technique, stacked plates or lamination techniques, fiber processing, and laser machining, have a more complex manufacturing process and introduce non-uniformity [75]; for instance, in the laser machining approach, rapid divergence of the tightly focused laser leads to thickness non-uniformity, this inhomogeneity causes interference in the signal generated from the traducer elements. In addition to the challenges in the fabrication process, incorporation of the backing layer in piezoelectric transducers adds manufacturing difficulty to maintain layer thickness uniformity. Since medium range frequencies are commonly used for PAI of biological tissues, the DF method can be utilized to manufacture piezoelectric transducer arrays.

3.2. Micromachined Ultrasonic Transducers

3.2.1. Capacitive Micromachined Ultrasonic Transducer (CMUT)

Capacitive micromachined ultrasonic transducers (CMUTs) are considered to be the next generation of ultrasound transducers [108]. CMUT is an array of miniaturized capacitors consisting of suspended membranes made of silicon nitride on dielectric posts, made of silicon nitride/oxide, with a conducting layer made of aluminum/gold and a rigid silicon conducting substrate as the base with a cavity in between. Different polymer materials (e.g., bisbenzocyclobutene) have also been used as the dielectric posts and diaphragms of CMUT arrays [109,110]. As opposed to the conventional piezoelectric transducers, CMUTs rely on electrostatic principles for ultrasound wave generation and reception when a superimposed DC bias and AC signal of desired frequency is applied [111] (see Figure 2a).

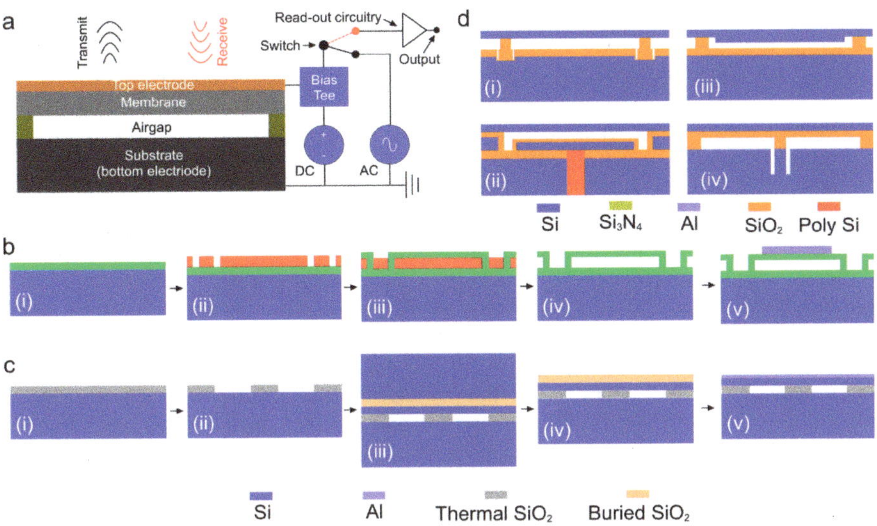

Figure 2. Capacitive micromachined ultrasonic transducer (CMUT) technology. (**a**) Schematic of a cross-section of a CMUT and its working principle, (**b**) steps of sacrificial release process: (i) substrate and insulation layer realization, (ii) sacrificial layer deposition and pattern, (iii) membrane layer deposition, (iv) sacrificial layer release, and (v) top electrode deposition, (**c**) steps of wafer bonding process: (i) thermal oxidation of silicon wafer (substrate), (ii) gap height and shape realization, (iii) bonding between silicon on insulator (SOI) and oxidized silicon wafer, (iv) thick silicon wafer etching, (v) buried oxide layer etching and top electrode realization [112], and (**d**) other CMUT designs fabricated using: (i) local oxidation of silicon (LOCOS) process [113], (ii) thick-buried-oxide process [114–117], (iii) mechanically coupled plate to the membrane [118], (iv) compliant post structure [114]. Reproduced with permission from [46].

Several process flows have been proposed by various research groups to implement CMUT arrays including, surface micromachining, fusion bonding, and adhesive bonding techniques [105]. Among those, the two most common fabrication methods of CMUT arrays are sacrificial release and wafer bonding processes. The basic process flow of the sacrificial release process is as follows (Figure 2b); initially, a sacrificial layer is deposited or grown on the carrier substrate. After membrane material deposition, the sacrificial layer is etched out with an etchant, specifically chosen for sacrificial layer material and not to etch the membrane layer material [115]. Although, sacrificial release process is relatively simple, reliable, and can be achieved at lower maximum processing temperature (250 °C) [116], non-uniform effective gap height due to the roughness in the silicon nitride layer causes deviations in device performance [117]. In addition, the diaphragm may induce substantial intrinsic stress that eventually alters the device properties. The basic process flow of the wafer bonding process is as follows (Figure 2c): initially, a highly doped silicon wafer is thermally oxidized to grow a SiO_2 layer followed by etching the oxide layer to determine the gap height and shape of the transducer elements. Next, silicon on insulator (SOI) wafer is brought in contact with the oxidized silicon wafer for bonding process [118]. The bulk silicon form SOI is removed by mechanical grinding and the buried oxide layer is etched to expose the Si diaphragm. Finally, a conducting layer such as aluminum is deposited for electrical routing. This process offers better control over gap height and thickness of the diaphragm with less residual stress. However, the wafer bonding process is very sensitive to surface roughness and cleanness that might affect the overall yield. There are a few other less popular fabrication processes such as local oxidation of silicon (LOCOS), thick buried oxide, mechanically coupled plate, and compliant post structure based CMUT arrays. The details of these processes can be found in [46]. A graphical representation of the final device structure for each method is depicted in Figure 2d.

CMUTs have gained much popularity over the last decade because they consume lower power, provide excellent electrical and thermal stability, and have a wider fractional bandwidth [46,104,105]. Micromachining techniques have advanced to allow batch fabrication of CMUT arrays of different shapes and frequencies on the same wafer with high yield and reduced price. CMUT technology has also enabled realizing densely packed elements in 2D configurations for volumetric imaging (see [119–121] for more details). The limitation of CMUT is that a large DC bias near the collapse voltage is required to achieve adequate sensitivity. This increases the risk of dielectric charging, changes the DC operating point which leads to an early breakdown of the device, hence greatly limiting the biomedical applicability of CMUT [111]. Hitachi Medico, Japan, Vermon, France, Butterfly Inc., USA, Kolo Medical, USA, Philips, USA, and Fraunhofer Institute for Photonic Microsystems (IPMS) are the pioneers in developing and commercializing CMUT technology.

As compared to conventional piezoelectric transducers, capacitive transducers may offer higher sensitivity and wider bandwidth as well as higher acceptance angle. These features are all important in photoacoustic imaging, where the spectral content of PA signals is distributed over a wide frequency range [43,122]. There is extensive literature discussing the use of CMUTs in different photoacoustic applications [43,119,120,123,124]. In Table 3, the existing CMUT probes that have been used in photoacoustic imaging applications are listed.

Table 3. Different configurations of CMUT that have been used in photoacoustic imaging applications.

Configuration	Element no.	CF (MHz)	BW (%)	Imaging Target	Ref
2D (16 × 16)	256	3.48	93.48	Fishing line filled with ICG, pig blood, and mixture of both	[120]
2D (16 × 16)	256	5	99	Tube filled with ink	[125]
2D (16 × 16)	256	5.5	112	Hair sample in tissue mimicking phantom	[119]
2D (Transparent)	NR	3.5	118	Wire phantom	[126]
2D (Transparent)	Single	1.46	105	Pencil lead; loop shaped tube filled with ICG	[126]
2D (Transparent)	NA	2	52.3	Characterization with hydrophone	[127]
Ring	NA	3	NA	Two polyethylene tubes	[128]
Hemisphere (spiral) *	500	4	>100	Arterioles and venules	[129]

BW: bandwidth; CF: center frequency, NA: not available. * Clinical application.

3.2.2. Piezoelectric Micromachined Ultrasonic Transducer (PMUT)

Piezoelectric micromachined ultrasound transducers (PMUTs) are low-cost technology with a high sensitivity that follows the principle of piezoelectric effect. In PMUT, an ultrasound wave is generated and detected based on flexural vibration of a diaphragm similar to a thin film on a silicon substrate without any vacuum gap [40] (see Figure 3a). Apart from the classification of sacrificial layer release and reverse wafer bonding methods that are similar to those used in manufacturing CMUT (see Figure 2b,c), there are two other methods to realize PMUT array diaphragms through back and front side etching (see Figure 3b,c) [130].

The manufacturing process of PMUT with circular diaphragms released from the front-side is described in [131]. For front-side etching (depicted in Figure 3b), a silicon wafer with a platinized thermal oxide layer is used as the substrate for the deposition of the piezoelectric and electrode layer. Lithography and reactive ion etch processes are used to pattern the top electrode, etch the thin film PZT layer to expose the bottom electrode followed by the deposition of insulation layer and electrode track fan-out to bonding pads. Then, the SiO_2/Ti/Pt/PZT/Pt thin film membranes are released from the Si substrate with XeF_2. Finally, the devices are laminated with a 15 μm thick dry film resist to seal the etched chambers and protect the thin film stack. In backside etching [132] (depicted in Figure 3c), the fabrication process flow starts with a Si (100) wafer [29]. As is the case for surface micromachining, the process begins with preparation of the insulator, (e.g., SiO_2 or Si_3N_4) on the silicon. This is then etched from one side of the Si in preparation for boron (B) doping. Boron diffusion occurs at a specific rate, allowing the control of the junction depth. After doping, the surface is cleaned and coated with low temperature oxide (LTO). Subsequently, standard photolithography is used to pattern the backside etch window. Later, the wafer is etched with an etchant such as ethylenediamine-pyrocatechol-water-pyrazine (EDP). After the back-side etching, a Ti/Pt bottom electrode is deposited by e-beam evaporation, followed by deposition of PZT and the top electrode. Finally, the top electrode and PZT are etched separately to pattern the top electrode and access the bottom electrode.

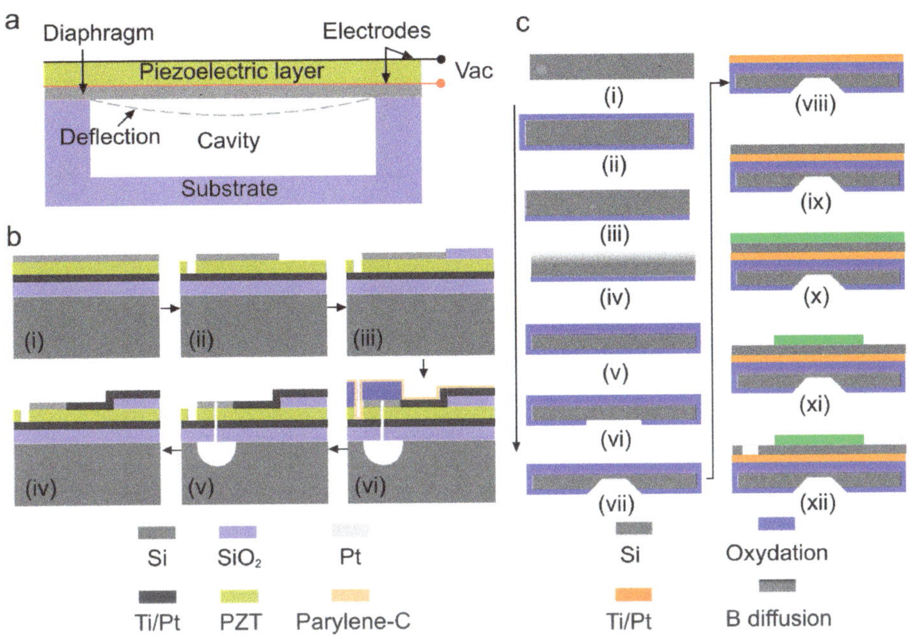

Figure 3. Piezoelectric micromachined ultrasound transducers (PMUT) technology. (**a**) Schematic of a cross-section of a PMUT and its working principle, (**b**) fabrication process flow of PMUTs with diaphragm defined by front-side etching method: (i) deposition of piezoelectric and electrode layer on top of oxidized silicon wafer, (i) SiO_2, Ti, Pt, and PZT layer grown on silicon wafer, (ii) pattern top electrode, (iii) insulation pad deposition, (iv) Ti/Pt deposition, (v) etch through layers to realize bias, and (vi) release the front side diaphragm, reprinted with permission from [130], and (**c**) steps of backside etching method: (i) silicon wafer, (ii) wet oxidation, (iii) oxide etching, (iv) boron diffusion, (v) low temperature oxide growth, (vi) oxide etch, (vii) Si etch, (viii) Ti-Pt deposition, (ix) PZT deposition, (x) TiW-Au deposition, (xi) top electrode etch, and (xii) PZT etch. Reprinted with permission from [132].

Since ultrasound transducers become smaller with increasing frequency, the effects of surface damage introduced during composite machining should be taken into account because the damaged layer volume increases in relation to the size of active piezoelectric materials. The use of a micromachining technique resolves the miniaturization issue of conventional piezoelectric transducers by realizing narrow channels or kerfs less than 10 microns, enabling high aspect ratio of piezoelectric elements [62]; this problem has been resolved in PMUT. Since the sensitivity is not limited unlike CMUTs, because they do not have a vacuum gap between the top and bottom electrodes, there is room to improve the coupling coefficient in PMUTs. Attributes of PMUTs such as low-cost with stable operation, established fabrication process, usage of popular materials (similar to conventional piezoelectric transducers) and capability of miniaturization [40,130–133] have made PMUTs a suitable candidate for photoacoustic imaging applications. Table 4 lists the studies where PMUTs have been used for photoacoustic imaging.

Table 4. Different configurations of PMUT that have been used in photoacoustic imaging applications.

Configuration	Element no.	CF (MHz)	BW (%)	Imaging Target	Ref
Linear	65	6.83	29.2	Six pencil leads at different depths	[134]
Linear	80	7	68%	Four pencil leads at different depths	[135]
1.5D Endoscopic	256 (32 × 16)	5	30	Metal spring; tricuspid valve and right ventricle in a porcine model	[136]

BW: bandwidth; CF: center frequency.

3.2.3. ASIC Technology in Physical Ultrasound Transducers

In clinical transducer arrays, each element is connected through a long wire to the analog-front-end (AFE) unit which includes transmit and receive beamformer, preamplifier, switches, and analog-digital converters (ADCs). Although this keeps all the electronics in one place, this arrangement causes interferences and reflections along the cable [137,138]. The number of cable connections can be reduced by multiplexing, however that has negative consequences such as limited bandwidth and slower processing [139]. CMOS technology-based application specific integrated circuits (ASIC) [140] is a novel technology, applicable to micromachined transducers, that is capable of integrating the AFE along with preamplifiers immediately after the ultrasound waves are received [141]. Philips, GE, and Siemens have successfully implemented ASICs within their probes (Philips X7-2t [142], GE 6VT-D [143], Siemens Z6M [144]). ASICs are also applicable to CMUTs and PMUTs [123,134]. Recently, Kolo Medical [145] and Butterfly Network [123] have launched commercial CMUT arrays based on SiliconWave™ and CMOS technologies, respectively. Since ASICs are custom designed, they are expensive, and their repair processes are still highly complicated.

3.3. Comparison between Physical Ultrasound Transducer Technologies

The physical ultrasound transducer technologies including PZT, CMUT, and PMUT are compared in terms of sensitivity, bandwidth, energy conversion and some other technical specifications in Table 5. Quantitative measurements of piezoelectric and CMUT are based on 2.43 and 2.63 MHz transducers, respectively, presented in [146,147] and that of PMUT are based on a 7~9 MHz transducer presented in [134,135].

Table 5. Comparison between physical ultrasound transducer technologies. DF: dice and fill, IC: integrated circuit, DC: direct current.

Parameters	Piezoelectric (PZT) [146,147]	CMUT [146,147]	PMUT [134,135]
Method	DF, Laminating	Wafer bonding, micromachining	Micromachining, wafer transfer
Sensitivity (mV/kPA)	4.28	22.57	0.48
Bandwidth (%)	60–80	≥100	50–60
Energy conversion (%)	45–75 [148]	>80	2.38–3.71
SNR (dB)	18–22 [149,150]	22–87 [120,151]	10–46
IC integration	Not compatible	Compatible	Compatible
Matching layer	Required	N/A	N/A
DC bias	N/A	Required	N/A

4. Optical Ultrasound Detection Technologies

The large size and optically opaque design of the widely used piezoelectric ultrasound transducers cause technical difficulties in some of the biomedical applications where optical illumination path and acoustic detection path must be coaxial. The mechanism that optical ultrasound detection methods offer could be a potential solution. This method employs high-finesse optical resonators to detect incident elastic waves. Providing miniaturized and optically transparent ultrasonic detectors [41], this technique yields a high sensitivity over a significantly wide frequency range, that are together ideal for photoacoustic imaging [42]. Wissmeyer et al. and Dong et al. have reviewed different methods with which optical ultrasound detection can be realized [41,42]. Based on different configurations and detection parameters, optical ultrasound detection techniques can be categorized into: (i) interferometric method and (ii) refractometric method [42]. Interferometric detection can be realized using Michelson interferometry (MI) [152,153], Mach-Zehnder interferometry (MZI) [154,155], doppler [156,157], or resonator [158–160]. In MI or MZI, two-beam method is employed where a laser beam passes into two optical paths, one of which is disturbed by the ultrasound wave and the other serves as a reference (see Figure 4a(i),(ii)). The changes in the optical path caused by the received pressure waves cause proportional changes in the intensity of the beam at the interferometer output [42]. In contrast to two-beam interferometers, doppler method senses ultrasound waves by measuring doppler shift (see Figure 4a(iii)). In resonator-based technique, a micron-scale optical resonator detects ultrasound waves (see Figure 4a(iv)); using this technique, miniaturization of the ultrasound detection unit is feasible. The optical resonator geometries that are most frequently used in photoacoustic imaging are Fabry–Pérot interferometers (FP) [161–163], micro-ring resonators (MRRs) [164–166], and π-phase-shifted fiber Bragg gratings (π-FBGs) [167–169]. Refractometric methods can be classified as intensity sensitive, beam deflectometry, and phase sensitive [41,42]. In intensity sensitive method [170,171], when the ultrasound waves pass through the interface of two media with different refractive indices, the intensity of the beam incident on that interface varies (see Figure 4b(i)). In the beam deflectometry method (see Figure 4b(ii)) [172,173], the interaction of the received acoustic waves with the medium alters the refractive index of the medium, which in turn deflects the probe beam that is eventually detected using a position-sensitive detector such as a quadrant photodiode [42]. In phase sensitive method [174], a collimated light beam passes through an acoustic field; the beam is deflected from the original path and perturbed; this beam is then focused through a spatial filter (see Figure 4b(iii)); the resultant beam is collimated and detected by a charge coupled device (CCD) or complementary metal-oxide-semiconductor (CMOS) camera; the image produced by the camera is the intensity map of the acoustic field.

One of the major limitations of optical ultrasound detection techniques is that they are slow. Although the scanning time can be reduced by parallelization [42,175], this would increase both the complexity and cost of the detection unit [41]. Another limitation is that these configurations mostly rely on continuous-wave (CW) lasers. CW interferometry is sensitive to temperature drifts and vibrations [42]. More details about the limitations of optical ultrasound detection techniques are given in [42,176–178]. Performance comparison between different optical ultrasound detectors is summarized in Table 6. The sensitivity of the optical ultrasound detection methods are represented in terms of noise equivalent pressure (NEP) that is a function of frequency [42]. By multiplying the square root of the center frequency, NEP can be presented in terms of pressure unit (Pascal) as shown in [41]; we used this unit in Table 6.

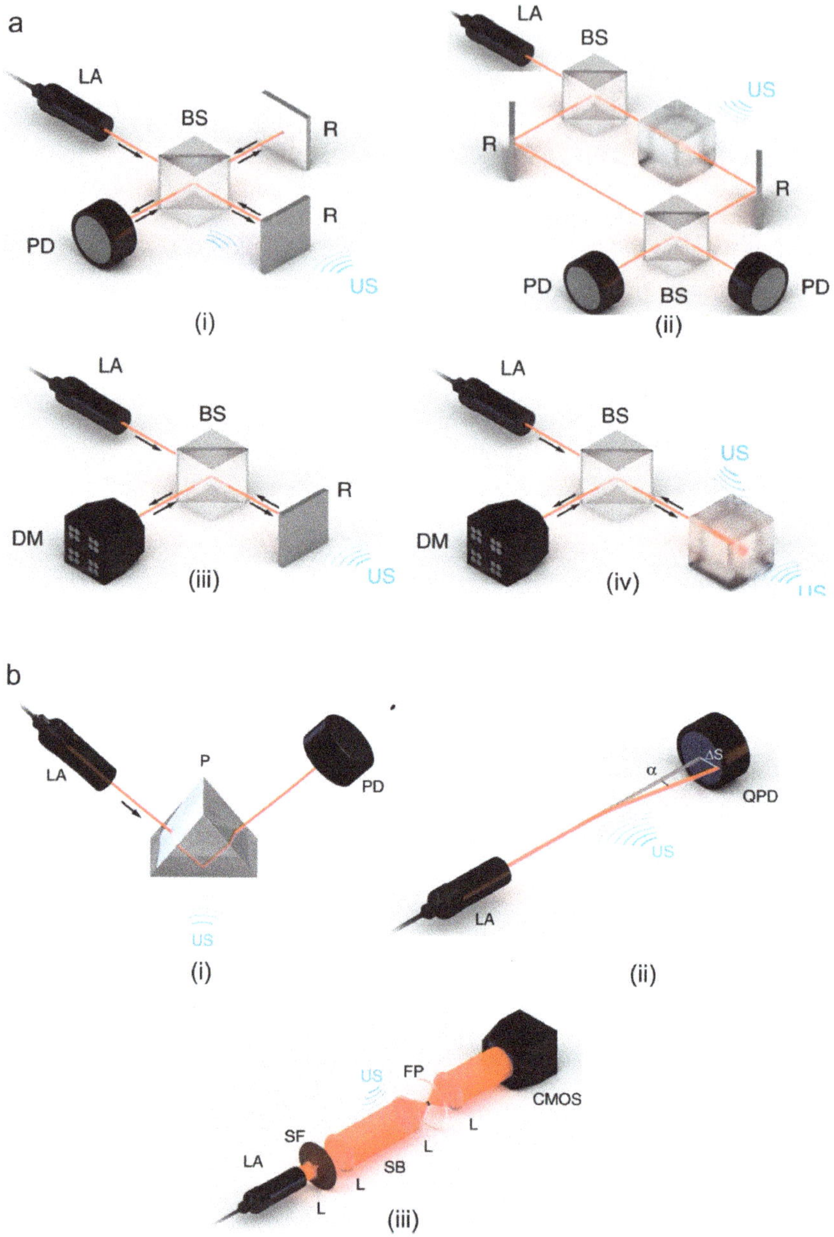

Figure 4. Optical ultrasound detection techniques. (**a**) Interferometric methods: (i) Michelson, (ii) Mach–Zehnder, (iii) doppler-based sensing, (iv) resonator-based sensing, and (**b**) refractometric: (i) intensity-sensitive detection of refractive index, (ii) single-beam deflectometry, (iii) phase-sensitive ultrasound detection. AL: acoustic lens, US: ultrasound, BS: beam splitter, D: detector, DM: demodulator, LA: laser, R: reflector, US: ultrasound, CMOS: complementary metal-oxide-semiconductor, FP: Fourier plane, L: lens, P: prism, PD: photodiode, QPD: quadrant photodiode, SB: Schlieren beam, SF: spatial filter. Reprinted with permission from [42].

Table 6. Summary of the performances of different optical ultrasound detection techniques. Reproduced from [41,42].

Method	Configuration		Readout Element Diameter/Dimension (µm)	Detection Geometry	BW (MHz)	NEP (Pa)	Ref.
Interferometric	MI		**	Point; disk	5; 20	35; 275	[179]
	MZI (free space)		90	Bar	17.5	100 (x mm)	[176]
	MZI (fiber optic)		125/8	Bar	50	92×10^3 (x mm)	[180]
	Doppler		-/12	Point	10	-	[181]
	RI	FP (free space)	**	Bar	25	20	[182]
		FP (fiber optic)	125/8	Bar	50	1	[180]
		MRR (integrated)	$60/0.8 \times 0.8$	Ring	140	6.8	[183]
		FBG (fiber optic)	$125/8 \times 100$	Bar	20	450	[184]
		FBG (integrated)	$500/1.5 \times 1.5$	Bar	60	6.5×10^3	[185]
Refractometric	Intensity-sensitive		15×10^{-3}	Prism	100	100 *	[186]
	Deflectometry		90	Needle beam	17	2.76 *	[172]
	Phase-sensitive		10^{-2}	Schlieren	110	486 *	[174]

* Unit: mPA·Hz$^{-1/2}$, ** Diffraction limited, BW: bandwidth; NEP: noise equivalent pressure, MI: Michelson interferometry; MZI: Mach–Zehnder interferometry; DI: doppler- interferometry; RI: resonator- interferometry.

5. Discussion and Conclusions

During the past several years, photoacoustic imaging technology has advanced in preclinical and clinical applications [7,8,16,18,122,187–193]. The clinical translation of this emerging imaging technology largely depends on the future of laser technology, data acquisition systems, and ultrasound transducer technology [194]. The ideal fabrication flow of a transducer device is as follows: depending on the application requirements such as geometrical restriction, desired penetration depth, and spatial resolution, the type and technology of the transducers are determined; an optimized structural/material design is then obtained by adjusting the geometric characteristics of the transducers' layers and their material properties; finally the transducers are built with a particular fabrication method, complexity of which depends on the budget. Ultrasound transducers with a high sensitivity in a wide spectral bandwidth, if cost is not a deciding factor, are ideal for photoacoustic imaging; a higher sensitivity can help reduce the necessary optical excitation energy and improve the penetration depth.

Among various ultrasound detection technologies, piezoelectric transducers are the most commonly used [195]; they have been made in forms of single element, as well as linear, arc, ring, hemispheric, and 2D matrix arrays. Their main limitations are that they require a matching layer, thermal instability, difficulty in realizing high frequency transducer arrays, and difficulty in miniaturization. As compared to piezoelectric transducers, CMUTs offer a higher sensitivity and a wider bandwidth, as well as a higher acceptance angle that are all important in photoacoustic imaging, where the spectral content of the PA signal is distributed over a wide frequency range [43,122]. In addition, fabrication of miniaturized transparent transducer arrays with desired shape is feasible using CMUTs. PMUT is a more recent technology with an improved bandwidth, higher sensitivity, lower acoustic impedance mismatch, flexible geometry, and the capability of CMOS/ASIC integration. PMUT, although does not outperform CMUT, relies on the established and reliable piezoelectric technology with micromachining capability. There is extensive research focused on improving the performance of PMUT that has led to promising results [40,130,133], therefore, despite better performance of CMUT, PMUT may have faster growth due to existing infrastructure.

In comparison with physical transducers, optical ultrasound detection technologies offer higher sensitivity over a significantly wide frequency range [42]. These technologies also demonstrate the capability of miniaturization and optically transparent transducers, which are both valuable features in biomedical imaging applications where optical illumination and acoustic detection paths must be coaxial for higher efficiency; this is an ideal arrangement of illumination and detection units in a photoacoustic imaging system. The main limitations of optical transducers are high sensitivity to temperature fluctuations and vibrations, as well as system cost. Overall, NEP of the reviewed technologies, (piezoelectric: 2 mPA·Hz$^{-1/2}$ [52], CMUT: 1.8~2.3 mPA·Hz$^{-1/2}$ [123], PMUT: 0.84~1.3 mPA·Hz$^{-1/2}$ [196], optical ultrasound detection: 0.45~486 mPA·Hz$^{-1/2}$ [42]) suggest that micromachined transducers (i.e., CMUTs and PMUTs) may be the more suitable transducers for photoacoustic imaging applications.

ASIC is a complimentary technology in transducer manufacturing to integrate the analog-front-end within the probe housing in order to reduce the noise of the transducer signal. ASIC improves the overall performance of existing transducers and therefore could help in facilitating the clinical translation of photoacoustic imaging. According to the technology market analyst projection, the usage of ASIC integrated miniature ultrasound transducer probes based on micromachined technologies will see ~18% annual compound growth rate by 2023 due to the advent of micromachining processes [197]. With the fast-growing ultrasound transducer technology, numerous computational methods have also been studied to further improve the performance of transducers by reducing noise in the transducer signal [198,199]; the result has been higher quality images [5,200,201]. With the advancement of ultrasound technology, more biomedical applications, which are currently performed using optical technologies, can be realized [12,202–208].

Author Contributions: Conceptualization, R.M. and K.A.; Methodology, K.A.; Software, R.M.; Validation, R.M., K.K., and K.A.; Formal Analysis, R.M. and K.A.; Investigation, R.M. and K.A.; Resources, K.A.; Writing-Original Draft Preparation, R.M.; Writing-Review & Editing, R.M., K.K., and K.A.; Visualization, R.M.; Supervision, K.A.; Funding Acquisition, K.A. All authors have read and agreed to the published version of the manuscript.

Funding: This work was supported by the National Institutes of Health R01EB027769-01 and R01EB028661-01.

Conflicts of Interest: The authors declare no conflict of interest.

References

1. Dance, D.R.; Christofides, S.; Maidment, A.; McLean, I.; Ng, K. *Diagnostic Radiology Physics: A Handbook for Teachers and Students. Endorsed by: American Association of Physicists in Medicine, Asia-Oceania Federation of Organizations for Medical Physics, European Federation of Organisations for Medical Physics;* International Atomic Energy Agency (IAEA): Vienna, Austria, 2014.
2. Wang, L.V. Tutorial on photoacoustic microscopy and computed tomography. *IEEE J. Sel. Top. Quantum Electron.* **2008**, *14*, 171–179. [CrossRef]
3. Wang, L.V.; Hu, S. Photoacoustic tomography: In vivo imaging from organelles to organs. *Science* **2012**, *335*, 1458–1462. [CrossRef] [PubMed]
4. Xia, J.; Yao, J.; Wang, L.V. Photoacoustic tomography: Principles and advances. *Electromagn. Waves (Camb.)* **2014**, *147*, 1–22. [CrossRef] [PubMed]
5. Yao, J.; Wang, L.V. Photoacoustic microscopy. *Laser Photonics Rev.* **2013**, *7*, 758–778. [CrossRef] [PubMed]
6. Zhou, Y.; Yao, J.; Wang, L.V. Tutorial on photoacoustic tomography. *J. Biomed. Opt.* **2016**, *21*, 061007. [CrossRef]
7. Manwar, R.; Hosseinzadeh, M.; Hariri, A.; Kratkiewicz, K.; Noei, S.; Avanaki, M.N. Photoacoustic signal enhancement: Towards Utilization of low energy laser diodes in real-time photoacoustic imaging. *Sensors* **2018**, *18*, 3498. [CrossRef]
8. Zafar, M.; Kratkiewicz, K.; Manwar, R.; Avanaki, M. Development of low-cost fast photoacoustic computed tomography: System characterization and phantom study. *Appl. Sci.* **2019**, *9*, 374. [CrossRef]
9. Fatima, A.; Kratkiewicz, K.; Manwar, R.; Zafar, M.; Zhang, R.; Huang, B.; Dadashzadesh, N.; Xia, J.; Avanaki, M. Review of cost reduction methods in photoacoustic computed tomography. *Photoacoustics* **2019**, *15*, 100137. [CrossRef]
10. Wang, L.V. *Photoacoustic Imaging and Spectroscopy*; CRC Press: Boca Raton, FL, USA, 2009.
11. Wang, L.V.; Yao, J. A practical guide to photoacoustic tomography in the life sciences. *Nat. Methods* **2016**, *13*, 627. [CrossRef]
12. Turani, Z.; Fatemizadeh, E.; Blumetti, T.; Daveluy, S.; Moraes, A.F.; Chen, W.; Mehregan, D.; Andersen, P.E.; Nasiriavanaki, M. Optical radiomic signatures derived from optical coherence tomography images improve identification of melanoma. *Cancer Res.* **2019**, *79*, 2021–2030. [CrossRef]
13. Rayyan, A.; Karl, K.; Mohammad, R.N.A. Photoacoustic imaging: A promising alternative to transcranial ultrasound. *Editor. Res. J. Photonics* **2018**, *2*, 411–420.

14. Zafar, M.; Kratkiewicz, K.; Manwar, R.; Avanaki, M. Low-cost fast photoacoustic computed tomography: Phantom study. In Proceedings of the Photons Plus Ultrasound: Imaging and Sensing 2019, San Francisco, CA, USA, 3–6 February 2019.
15. Mozaffarzadeh, M.; Mahloojifar, A.; Orooji, M.; Adabi, S.; Nasiriavanaki, M. Double-stage delay multiply and sum beamforming algorithm: Application to linear-array photoacoustic imaging. *IEEE Trans. Biomed. Eng.* **2018**, *65*, 31–42. [CrossRef]
16. Mozaffarzadeh, M.; Mahloojifar, A.; Orooji, M.; Kratkiewicz, K.; Adabi, S.; Nasiriavanaki, M. Linear-array photoacoustic imaging using minimum variance-based delay multiply and sum adaptive beamforming algorithm. *J. Biomed. Opt.* **2018**, *23*, 026002. [CrossRef] [PubMed]
17. Omidi, P.; Zafar, M.; Mozaffarzadeh, M.; Hariri, A.; Haung, X.; Orooji, M.; Nasiriavanaki, M. A novel dictionary-based image reconstruction for photoacoustic computed tomography. *Appl. Sci.* **2018**, *8*, 1570. [CrossRef]
18. Nasiriavanaki, M.; Xia, J.; Wan, H.; Bauer, A.Q.; Culver, J.P.; Wang, L.V. High-resolution photoacoustic tomography of resting-state functional connectivity in the mouse brain. *Proc. Natl. Acad. Sci. USA* **2014**, *111*, 21–26. [CrossRef] [PubMed]
19. Upputuri, P.K.; Periyasamy, V.; Kalva, S.K.; Pramanik, M. A High-performance Compact Photoacoustic Tomography System for In Vivo Small-animal Brain Imaging. *J. Vis. Exp.* **2017**, *124*, e55811. [CrossRef]
20. Xia, J.; Wang, L.V. Small-Animal Whole-Body Photoacoustic Tomography: A Review. *IEEE Trans. Biomed. Eng.* **2014**, *61*, 1380–1389. [CrossRef] [PubMed]
21. Upputuri, P.K.; Pramanik, M. Dynamic in vivo imaging of small animal brain using pulsed laser diode-based photoacoustic tomography system. *J. Biomed. Opt.* **2017**, *22*, 1. [CrossRef] [PubMed]
22. Upputuri, P.K.; Pramanik, M. Photoacoustic imaging in the second near-infrared window: A review. *J. Biomed. Opt.* **2019**, *24*, 1–20. [CrossRef]
23. Liu, W.; Zhang, H.F. Photoacoustic imaging of the eye: A mini review. *Photoacoustics* **2016**, *4*, 112–123. [CrossRef]
24. Silverman, R.H.; Kong, F.; Chen, Y.; Lloyd, H.O.; Kim, H.H.; Cannata, J.M.; Shung, K.K.; Coleman, D.J. High-Resolution Photoacoustic Imaging of Ocular Tissues. *Ultrasound Med. Boil.* **2010**, *36*, 733–742. [CrossRef] [PubMed]
25. Nguyen, V.P.; Paulus, Y.M. Photoacoustic Ophthalmoscopy: Principle, Application, and Future Directions. *J. Imaging* **2018**, *4*, 149. [CrossRef] [PubMed]
26. Jeon, S.; Song, H.B.; Kim, J.; Lee, B.J.; Managuli, R.; Kim, J.H.; Kim, J.H.; Kim, C. In Vivo Photoacoustic Imaging of Anterior Ocular Vasculature: A Random Sample Consensus Approach. *Sci. Rep.* **2017**, *7*, 4318. [CrossRef]
27. Wang, S.; Lin, J.; Wang, T.; Chen, X.; Huang, P. Recent Advances in Photoacoustic Imaging for Deep-Tissue Biomedical Applications. *Theranostics* **2016**, *6*, 2394–2413. [CrossRef]
28. Mehrmohammadi, M.; Yoon, S.J.; Yeager, D.; Emelianov, S.Y. Photoacoustic Imaging for Cancer Detection and Staging. *Curr. Mol. Imaging* **2013**, *2*, 89–105. [CrossRef] [PubMed]
29. Oh, J.-T.; Li, M.-L.; Zhang, H.F.; Maslov, K.; Stoica, G.; Wang, L.V. Three-dimensional imaging of skin melanoma in vivo by dual-wavelength photoacoustic microscopy. *J. Biomed. Opt.* **2006**, *11*, 034032. [CrossRef]
30. Rao, A.P.; Bokde, N.D.; Sinha, S. Photoacoustic Imaging for Management of Breast Cancer: A Literature Review and Future Perspectives. *Appl. Sci.* **2020**, *10*, 767. [CrossRef]
31. Steinberg, I.; Huland, D.M.; Vermesh, O.; Frostig, H.E.; Tummers, W.S.; Gambhir, S.S. Photoacoustic clinical imaging. *Photoacoustics* **2019**, *14*, 77–98. [CrossRef]
32. Fakhrejahani, E.; Torii, M.; Kitai, T.; Kanao, S.; Asao, Y.; Hashizume, Y.; Mikami, Y.; Yamaga, I.; Kataoka, M.; Sugie, T.; et al. Clinical Report on the First Prototype of a Photoacoustic Tomography System with Dual Illumination for Breast Cancer Imaging. *PLoS ONE* **2015**, *10*, e0139113. [CrossRef]
33. Brown, E.; Brunker, J.; Bohndiek, S. Photoacoustic imaging as a tool to probe the tumour microenvironment. *Dis. Model. Mech.* **2019**, *12*, dmm039636. [CrossRef]
34. Valluru, K.S.; Willmann, J.K. Clinical photoacoustic imaging of cancer. *Ultrason.* **2016**, *35*, 267–280. [CrossRef] [PubMed]

35. Zhang, C.; Zhang, Y.; Hong, K.; Zhu, S.; Wan, J. Photoacoustic and fluorescence imaging of cutaneous squamous cell carcinoma in living subjects using a probe targeting integrin $\alpha_v\beta_6$. *Sci. Rep.* **2017**, *7*, 42442. [CrossRef]
36. Zhang, E.Z.; Považay, B.; Laufer, J.; Alex, A.; Hofer, B.; Pedley, B.; Glittenberg, C.; Treeby, B.; Cox, B.T.; Beard, P.; et al. Multimodal photoacoustic and optical coherence tomography scanner using an all optical detection scheme for 3D morphological skin imaging. *Biomed. Opt. Express* **2011**, *2*, 2202–2215. [CrossRef] [PubMed]
37. Guo, B.; Sheng, Z.; Hu, D.; Liu, C.; Wang, X.; Liu, B. Through Scalp and Skull NIR-II Photothermal Therapy of Deep Orthotopic Brain Tumors with Precise Photoacoustic Imaging Guidance. *Adv. Mater.* **2018**, *30*, e1802591. [CrossRef] [PubMed]
38. Joshi, B.P.; Wang, T.D. Targeted Optical Imaging Agents in Cancer: Focus on Clinical Applications. *Contrast Media Mol. Imaging* **2018**, *2018*, 1–19. [CrossRef]
39. Lee, W.; Roh, Y. Ultrasonic transducers for medical diagnostic imaging. *Biomed. Eng. Lett.* **2017**, *7*, 91–97. [CrossRef]
40. Jung, J.; Lee, W.; Kang, W.; Shin, E.; Ryu, J.; Choi, H. Review of piezoelectric micromachined ultrasonic transducers and their applications. *J. Micromech. Microeng.* **2017**, *27*, 113001. [CrossRef]
41. Dong, B.; Sun, C.; Zhang, H.F. Optical detection of ultrasound in photoacoustic imaging. *IEEE Trans. Biomed. Eng.* **2017**, *64*, 4–15. [CrossRef]
42. Wissmeyer, G.; Pleitez, M.A.; Rosenthal, A.; Ntziachristos, V. Looking at sound: Optoacoustics with all-optical ultrasound detection. *Light Sci. Appl.* **2018**, *7*, 53. [CrossRef]
43. Vallet, M.; Varray, F.; Boutet, J.; Dinten, J.-M.; Caliano, G.; Savoia, A.S.; Vray, D. Quantitative comparison of PZT and CMUT probes for photoacoustic imaging: Experimental validation. *Photoacoustics* **2017**, *8*, 48–58. [CrossRef]
44. Jiang, X.; Al-Jumaily, A.M. Ultrasound transducers for biomedical imaging and therapy. *J. Eng. Sci. Med. Diagn. Ther.* **2018**, *1*, 040201. [CrossRef]
45. Salim, M.S.; Abd Malek, M.F.; Heng, R.B.W.; Juni, K.M.; Sabri, N. Capacitive micromachined ultrasonic transducers: Technology and application. *J. Med. Ultrasound* **2012**, *20*, 8–31. [CrossRef]
46. Khuri-Yakub, B.T.; Oralkan, O. Capacitive micromachined ultrasonic transducers for medical imaging and therapy. *J. Micromech. Microeng. Struct. Devices Syst.* **2011**, *21*, 54004–54014. [CrossRef] [PubMed]
47. Duy Le, A. *Investigation of Element Variations in Ultrasound Transducer Arrays by Electrical Impedance Measurements*; Høgskolen i Sørøst-Norge: Notodden, Norway, 2016.
48. Yaralioglu, G.G.; Ergun, A.S.; Bayram, B.; Haeggstrom, E.; Khuri-Yakub, B.T. Calculation and measurement of electromechanical coupling coefficient of capacitive micromachined ultrasonic transducers. *IEEE Trans. Ultrason. Ferroelectr. Freq. Control* **2003**, *50*, 449–456. [CrossRef] [PubMed]
49. Yang, Y.; Wei, X.; Zhang, L.; Yao, W. The effect of electrical impedance matching on the electromechanical characteristics of sandwiched piezoelectric ultrasonic transducers. *Sensors* **2017**, *17*, 2832. [CrossRef] [PubMed]
50. Rathod, V.T. A review of electric impedance matching techniques for piezoelectric sensors, actuators and transducers. *Electronics* **2019**, *8*, 169. [CrossRef]
51. Ritter, T.; Shung, K.; Hackenberger, W.; Wang, H.; Shrout, T.R. Performance of a high dielectric constant piezoelectric ceramic for ultrasound transducers. In Proceedings of the 1999 IEEE Ultrasonics Symposium. Proceedings. International Symposium (Cat. No. 99CH37027), Caesars Tahoe, NV, USA, 17–20 October 1999.
52. Winkler, A.M.; Maslov, K.I.; Wang, L.V. Noise-equivalent sensitivity of photoacoustics. *J. Biomed. Opt.* **2013**, *18*, 097003. [CrossRef]
53. Yao, J.; Wang, L.V. Sensitivity of photoacoustic microscopy. *Photoacoustics* **2014**, *2*, 87–101. [CrossRef]
54. Zhou, Q.; Lam, K.H.; Zheng, H.; Qiu, W.; Shung, K.K. Piezoelectric single crystal ultrasonic transducers for biomedical applications. *Prog. Mater. Sci.* **2014**, *66*, 87–111. [CrossRef] [PubMed]
55. Szabo, T.L.; Lewin, P.A. Ultrasound transducer selection in clinical imaging practice. *J. Ultrasound Med.* **2013**, *32*, 573–582. [CrossRef]
56. Zhang, J.X.J.; Hoshino, K. *Molecular Sensors and Nanodevices: Principles, Designs and Applications in Biomedical Engineering*; Academic Press: Cambridge, MA, USA, 2018.
57. Lerch, R. Finite element analysis of piezoelectric transducers. In Proceedings of the IEEE 1988 Ultrasonics Symposium Proceedings, Chicago, IL, USA, 2–5 October 1988.

58. Loveday, P.W. Analysis of piezoelectric ultrasonic transducers attached to waveguides using waveguide finite elements. *IEEE Trans. Ultrason. Ferroelectr. Freq. Control* **2007**, *54*, 2045–2051. [CrossRef] [PubMed]
59. Hepher, M.J. Natural and synthetic piezoelectric materials for chemical sensors. In Proceedings of the 1992 Sixth International Conference on Dielectric Materials, Measurements and Applications, Manchester, UK, 7–10 September 1992.
60. Kumar, P.; Singh, S.; Thakur, O.P.; Prakash, C.; Goel, T.C. Study of lead magnesium niobate–lead titanate ceramics for piezo-actuator applications. *Jpn. J. Appl. Phys.* **2004**, *43*, 1501. [CrossRef]
61. Ren, W.; Liu, S.-F.; Mukherjee, B.K. Nonlinear behavior of piezoelectric lead zinc niobate-lead titanate single crystals under AC electric fields. In Proceedings of the Smart Structures and Materials 2003: Active Materials: Behavior and Mechanics, San Diego, CA, USA, 3–6 March 2003.
62. Lee, H.J.; Zhang, S.; Bar-Cohen, Y.; Sherrit, S. High temperature, high power piezoelectric composite transducers. *Sensors* **2014**, *14*, 14526–14552. [CrossRef] [PubMed]
63. Huang, Q.; Wang, H.; Hao, S.; Zhong, C.; Wang, L. Design and fabrication of a high-frequency single-directional planar underwater ultrasound transducer. *Sensors* **2019**, *19*, 4336. [CrossRef]
64. Acosta, M.; Novak, N.; Rojas, V.; Patel, S.; Vaish, R.; Koruza, J.; Rossetti, G., Jr.; Rödel, J. $BaTiO_3$-based piezoelectrics: Fundamentals, current status, and perspectives. *Appl. Phys. Rev.* **2017**, *4*, 041305. [CrossRef]
65. Bordui, P.F.; Norwood, R.; Jundt, D.H.; Fejer, M. Preparation and characterization of off-congruent lithium niobate crystals. *J. Appl. Phys.* **1992**, *71*, 875–879. [CrossRef]
66. Yamada, T.; Niizeki, N.; Toyoda, H. Piezoelectric and elastic properties of lithium niobate single crystals. *Jpn. J. Appl. Phys.* **1967**, *6*, 151. [CrossRef]
67. Sengupta, P.; Khanra, K.; Chowdhury, A.R.; Datta, P. 4—Lab-on-a-chip sensing devices for biomedical applications. In *Bioelectronics and Medical Devices*; Pal, K., Ed.; Woodhead Publishing: Sawston, UK, 2019; pp. 47–95.
68. Denishev, K. Some metal oxides and their applications for creation of Microsystems (MEMS) and Energy Harvesting Devices (EHD). *J. Phys. Conf. Ser.* **2016**, *764*, 012003. [CrossRef]
69. Safari, A.; Akdogan, E.K. *Piezoelectric and Acoustic Materials for Transducer Applications*; Springer Science & Business Media: Berlin/Heidelberg, Germany, 2008.
70. Ketterling, J.A.; Aristizabal, O.; Turnbull, D.H.; Lizzi, F.L. Design and fabrication of a 40-MHz annular array transducer. *IEEE Trans. Ultrason. Ferroelectr. Freq. Control* **2005**, *52*, 672–681. [CrossRef]
71. Sherar, M.D.; Foster, F.S. The design and fabrication of high frequency poly (vinylidene fluoride) transducers. *Ultrason. Imaging* **1989**, *11*, 75–94. [CrossRef]
72. Sherar, M.D.; Foster, F.S. A 100 MHz PVDF ultrasound microscope with biological applications. In *Acoustical Imaging*; Springer: Berlin/Heidelberg, Germany, 1988; pp. 511–520.
73. Foster, F.S.; Harasiewicz, K.A.; Sherar, M.D. A history of medical and biological imaging with polyvinylidene fluoride (PVDF) transducers. *IEEE Trans. Ultrason. Ferroelectr. Freq. Control* **2000**, *47*, 1363–1371. [CrossRef]
74. Decharat, A.; Wagle, S.; Melandsø, F. Effect of polymer electrode thickness on the acoustical properties of all-screen printed piezoelectric pvdf copolymer transducers. *Jpn. J. Appl. Phys.* **2014**, *53*, 05HB16. [CrossRef]
75. Safari, A. Development of piezoelectric composites for transducers. *J. Phys. III* **1994**, *4*, 1129–1149. [CrossRef]
76. Sun, R.; Wang, L.; Zhang, Y.; Zhong, C. Characterization of 1-3 piezoelectric composite with a 3-tier polymer structure. *Materials* **2020**, *13*, 397. [CrossRef] [PubMed]
77. Goldberg, R.L.; Smith, S.W. Multilayer piezoelectric ceramics for two-dimensional array transducers. *IEEE Trans. Ultrason. Ferroelectr. Freq. Control* **1994**, *41*, 761–771. [CrossRef]
78. Martin, K.H.; Lindsey, B.D.; Ma, J.; Lee, M.; Li, S.; Foster, F.S.; Jiang, X.; Dayton, P.A. Dual-frequency piezoelectric transducers for contrast enhanced ultrasound imaging. *Sensors* **2014**, *14*, 20825–20842. [CrossRef]
79. Muralt, P.; Ledermann, N.; Paborowski, J.; Barzegar, A.; Gentil, S.; Belgacem, B.; Petitgrand, S.; Bosseboeuf, A.; Setter, N. Piezoelectric micromachined ultrasonic transducers based on PZT thin films. *IEEE Trans. Ultrason. Ferroelectr. Freq. Control* **2005**, *52*, 2276–2288. [CrossRef]
80. Zipparo, M.J.; Shung, K.K.; Shrout, T.R. Piezoceramics for high-frequency (20 to 100 MHz) single-element imaging transducers. *IEEE Trans. Ultrason. Ferroelectr. Freq. Control* **1997**, *44*, 1038–1048. [CrossRef]
81. Newnham, R.E.; Skinner, D.P.; Cross, L.E. Connectivity and piezoelectric-pyroelectric composites. *Mater. Res. Bull.* **1978**, *13*, 525–536. [CrossRef]
82. Cross, L.E. Ferroelectric materials for electromechanical transducer applications. *Mater. Chem. Phys.* **1996**, *43*, 108–115. [CrossRef]

83. Silva, E.C.N.; Fonseca, J.O.; de Espinosa, F.M.; Crumm, A.; Brady, G.; Halloran, J.; Kikuchi, N. Design of piezocomposite materials and piezoelectric transducers using topology optimization—Part I. *Arch. Comput. Methods Eng.* **1999**, *6*, 117–182. [CrossRef]
84. Levassort, F.; Tran-Huu-Hue, L.P.; Certon, D.; Lethiecq, M. *Piezoelectric Materials for Ultrasonic Transducers: Review of Recent Developments*; Laboratoire d'Ultrasons, Signaux et Instrumentation, GIP Ultrasons: Blois, France, 2002.
85. Hackenberger, W.; Rehrig, P.W.; Ritter, T.; Shrout, T. Advanced piezoelectric materials for medical ultrasound transducers. In Proceedings of the 2001 IEEE Ultrasonics Symposium. Proceedings. An International Symposium (Cat. No. 01CH37263), Atlanta, GA, USA, 7–10 October 2001.
86. Thippabhotla, S.; Zhong, C.; He, M. 3D cell culture stimulates the secretion of in vivo like extracellular vesicles. *Sci. Rep.* **2019**, *9*, 13012. [CrossRef] [PubMed]
87. Heijblom, M.; Piras, D.; van den Engh, F.M.; van der Schaaf, M.; Klaase, J.M.; Steenbergen, W.; Manohar, S. The state of the art in breast imaging using the Twente Photoacoustic Mammoscope: Results from 31 measurements on malignancies. *Eur. Radiol.* **2016**, *26*, 3874–3887. [CrossRef]
88. Toi, M.; Asao, Y.; Matsumoto, Y.; Sekiguchi, H.; Yoshikawa, A.; Takada, M.; Kataoka, M.; Endo, T.; Kawaguchi-Sakita, N.; Kawashima, M. Visualization of tumor-related blood vessels in human breast by photoacoustic imaging system with a hemispherical detector array. *Sci. Rep.* **2017**, *7*, 41970. [CrossRef]
89. Ermilov, S.A.; Khamapirad, T.; Conjusteau, A.; Leonard, M.H.; Lacewell, R.; Mehta, K.; Miller, T.; Oraevsky, A.A. Laser optoacoustic imaging system for detection of breast cancer. *J. Biomed. Opt.* **2009**, *14*, 024007. [CrossRef] [PubMed]
90. Schwarz, M.; Buehler, A.; Aguirre, J.; Ntziachristos, V. Three-dimensional multispectral optoacoustic mesoscopy reveals melanin and blood oxygenation in human skin in vivo. *J. Biophotonics* **2016**, *9*, 55–60. [CrossRef] [PubMed]
91. Aguirre, J.; Schwarz, M.; Soliman, D.; Buehler, A.; Omar, M.; Ntziachristos, V. Broadband mesoscopic optoacoustic tomography reveals skin layers. *Opt. Lett.* **2014**, *39*, 6297–6300. [CrossRef]
92. Petri, M.; Stoffels, I.; Jose, J.; Leyh, J.; Schulz, A.; Dissemond, J.; Schadendorf, D.; Klode, J. Photoacoustic imaging of real-time oxygen changes in chronic leg ulcers after topical application of a haemoglobin spray: A pilot study. *J. Wound Care* **2016**, *25*, 87–89. [CrossRef]
93. Favazza, C.P.; Jassim, O.; Cornelius, L.A.; Wang, L.V. In vivo photoacoustic microscopy of human cutaneous microvasculature and a nevus. *J. Biomed. Opt.* **2011**, *16*, 016015. [CrossRef]
94. Dima, A.; Ntziachristos, V. Non-invasive carotid imaging using optoacoustic tomography. *Opt. Express* **2012**, *20*, 25044–25057. [CrossRef]
95. Seeger, M.; Karlas, A.; Soliman, D.; Pelisek, J.; Ntziachristos, V. Multimodal optoacoustic and multiphoton microscopy of human carotid atheroma. *Photoacoustics* **2016**, *4*, 102–111. [CrossRef]
96. Van Es, P.; Biswas, S.K.; Moens, H.J.B.; Steenbergen, W.; Manohar, S. Initial results of finger imaging using photoacoustic computed tomography. *J. Biomed. Opt.* **2014**, *19*, 060501. [CrossRef] [PubMed]
97. Yuan, J.; Xu, G.; Yu, Y.; Zhou, Y.; Carson, P.L.; Wang, X.; Liu, X. Real-time photoacoustic and ultrasound dual-modality imaging system facilitated with graphics processing unit and code parallel optimization. *J. Biomed. Opt.* **2013**, *18*, 086001. [CrossRef] [PubMed]
98. Buehler, A.; Diot, G.; Volz, T.; Kohlmeyer, J.; Ntziachristos, V. Imaging of fatty tumors: Appearance of subcutaneous lipomas in optoacoustic images. *J. Biophotonics* **2017**, *10*, 983–989. [CrossRef] [PubMed]
99. Reber, J.; Willershäuser, M.; Karlas, A.; Paul-Yuan, K.; Diot, G.; Franz, D.; Fromme, T.; Ovsepian, S.V.; Bézière, N.; Dubikovskaya, E. Non-invasive measurement of brown fat metabolism based on optoacoustic imaging of hemoglobin gradients. *Cell Metab.* **2018**, *27*, 689–701. [CrossRef] [PubMed]
100. Yang, M.; Zhao, L.; He, X.; Su, N.; Zhao, C.; Tang, H.; Hong, T.; Li, W.; Yang, F.; Lin, L. Photoacoustic/ultrasound dual imaging of human thyroid cancers: An initial clinical study. *Biomed. Opt. Express* **2017**, *8*, 3449–3457. [CrossRef] [PubMed]
101. Dima, A.; Ntziachristos, V. In-vivo handheld optoacoustic tomography of the human thyroid. *Photoacoustics* **2016**, *4*, 65–69. [CrossRef] [PubMed]
102. Nedosekin, D.A.; Sarimollaoglu, M.; Ye, J.H.; Galanzha, E.I.; Zharov, V.P. In vivo ultra-fast photoacoustic flow cytometry of circulating human melanoma cells using near-infrared high-pulse rate lasers. *Cytom. Part A* **2011**, *79*, 825–833. [CrossRef]

103. Horiguchi, A.; Tsujita, K.; Irisawa, K.; Kasamatsu, T.; Hirota, K.; Kawaguchi, M.; Shinchi, M.; Ito, K.; Asano, T.; Shinmoto, H. A pilot study of photoacoustic imaging system for improved real-time visualization of neurovascular bundle during radical prostatectomy. *Prostate* **2016**, *76*, 307–315. [CrossRef]
104. Manwar, R. A BCB Diaphragm Based Adhesive Wafer Bonded CMUT Probe for Biomedical Application. Ph.D. Thesis, University of Windsor, Windsor, ON, Canada, 2017.
105. Leondes, C.T. *Mems/Nems:(1) Handbook Techniques and Applications Design Methods,(2) Fabrication Techniques,(3) Manufacturing Methods,(4) Sensors and Actuators,(5) Medical Applications and MOEMS*; Springer Science & Business Media: Berlin/Heidelberg, Germany, 2007.
106. Soumia, E.H.; Mohamad, E. *Recent Advances in Electrical and Information Technologies for Sustainable Development*; Springer International Publishing: Cham, Switzerland, 2019.
107. Liu, C.; Djuth, F.T.; Zhou, Q.; Shung, K.K. Micromachining techniques in developing high-frequency piezoelectric composite ultrasonic array transducers. *IEEE Trans. Ultrason. Ferroelectr. Freq. Control* **2013**, *60*, 2615–2625.
108. Oralkan, O.; Ergun, A.S.; Johnson, J.A.; Karaman, M.; Demirci, U.; Kaviani, K.; Lee, T.H.; Khuri-Yakub, B.T. Capacitive micromachined ultrasonic transducers: Next-generation arrays for acoustic imaging? *IEEE Trans. Ultrason. Ferroelectr. Freq. Control* **2002**, *49*, 1596–1610. [CrossRef]
109. Manwar, R.; Chowdhury, S. Experimental analysis of bisbenzocyclobutene bonded capacitive micromachined ultrasonic transducers. *Sensors* **2016**, *16*, 959. [CrossRef] [PubMed]
110. Li, Z.; Wong, L.L.; Chen, A.I.; Na, S.; Sun, J.; Yeow, J.T. Fabrication of capacitive micromachined ultrasonic transducers based on adhesive wafer bonding technique. *J. Micromech. Microeng.* **2016**, *26*, 115019. [CrossRef]
111. Manwar, R.; Simpson, T.; Bakhtazad, A.; Chowdhury, S. Fabrication and characterization of a high frequency and high coupling coefficient CMUT array. *Microsyst. Technol.* **2017**, *23*, 4965–4977. [CrossRef]
112. Kupnik, M.; Vaithilingam, S.; Torashima, K.; Wygant, I.O.; Khuri-Yakub, B.T. CMUT fabrication based on a thick buried oxide layer. In Proceedings of the 2010 IEEE International Ultrasonics Symposium, San Diego, CA, USA, 11–14 October 2010.
113. Huang, Y.; Zhuang, X.; Haeggstrom, E.O.; Ergun, A.S.; Cheng, C.H.; Khuri-Yakub, B.T. Capacitive micromachined ultrasonic transducers with piston-shaped membranes: Fabrication and experimental characterization. *IEEE Trans. Ultrason. Ferroelectr. Freq. Control* **2009**, *56*, 136–145. [CrossRef] [PubMed]
114. Nikoozadeh, A.; Khuri-Yakub, P.T. CMUT with substrate-embedded springs for non-flexural plate movement. In Proceedings of the IEEE Ultrasonics Symposium, San Diego, CA, USA, 11–14 October 2010; pp. 1510–1513.
115. Erguri, A.S.; Huang, Y.; Zhuang, X.; Oralkan, O.; Yarahoglu, G.G.; Khuri-Yakub, B.T. Capacitive micromachined ultrasonic transducers: Fabrication technology. *IEEE Trans. Ultrason. Ferroelectr. Freq. Control* **2005**, *52*, 2242–2258. [CrossRef]
116. Knight, J.; McLean, J.; Degertekin, F.L. Low temperature fabrication of immersion capacitive micromachined ultrasonic transducers on silicon and dielectric substrates. *IEEE Trans. Ultrason. Ferroelectr. Freq. Control* **2004**, *51*, 1324–1333. [CrossRef]
117. Lin, D.; Zhuang, X.; Wong, S.H.; Ergun, A.S.; Kupnik, M.; Khuri-Yakub, B.T. 6F-5 Characterization of Fabrication Related Gap-Height Variations in Capacitive Micromachined Ultrasonic Transducers. In Proceedings of the 2007 IEEE Ultrasonics Symposium Proceedings, New York, NY, USA, 28–31 October 2007.
118. Brenner, K.; Ergun, A.S.; Firouzi, K.; Rasmussen, M.F.; Stedman, Q.; Khuri-Yakub, B.P. Advances in capacitive micromachined ultrasonic transducers. *Micromachines* **2019**, *10*, 152. [CrossRef] [PubMed]
119. Kothapalli, S.; Ma, T.-J.; Vaithilingam, S.; Oralkan, Ö.; Khuri-Yakub, B.T.; Gambhir, S.S. Deep tissue photoacoustic imaging using a miniaturized 2-D capacitive micromachined ultrasonic transducer array. *IEEE Trans. Biomed. Eng.* **2012**, *59*, 1199–1204. [CrossRef]
120. Wygant, I.O.; Zhuang, X.; Kuo, P.; Yeh, D.; Oralkan, O.; Khuri-Yakub, B. Photoacoustic imaging using a two-dimensional CMUT array. In Proceedings of the IEEE Ultrasonics Symposium, Rotterdam, The Netherlands, 18–21 September 2005.
121. Vaithilingam, S.; Ma, T.-J.; Furukawa, Y.; Wygant, I.O.; Zhuang, X.; De La Zerda, A.; Oralkan, O.; Kamaya, A.; Jeffrey, R.B.; Khuri-Yakub, B.T. Three-dimensional photoacoustic imaging using a two-dimensional CMUT array. *IEEE Trans. Ultrason. Ferroelectr. Freq. Control* **2009**, *56*, 2411–2419. [CrossRef]

122. Mahmoodkalayeh, S.; Jooya, H.Z.; Hariri, A.; Zhou, Y.; Xu, Q.; Ansari, M.A.; Avanaki, M.R. Low temperature-mediated enhancement of photoacoustic imaging depth. *Sci. Rep.* **2018**, *8*, 4873. [CrossRef]
123. Wygant, I.O.; Zhuang, X.; Yeh, D.T.; Oralkan, O.; Ergun, A.S.; Karaman, M.; Khuri-Yakub, B.T. Integration of 2D CMUT arrays with front-end electronics for volumetric ultrasound imaging. *IEEE Trans. Ultrason. Ferroelectr. Freq. Control* **2008**, *55*, 327–342. [CrossRef] [PubMed]
124. Chee, R.K.W.; Sampaleanu, A.; Rishi, D.; Zemp, R.J. Top orthogonal to bottom electrode (TOBE) 2-D CMUT arrays for 3-D photoacoustic imaging. *IEEE Trans. Ultrason. Ferroelectr. Freq. Control* **2014**, *61*, 1393–1395. [CrossRef] [PubMed]
125. Ma, T.-J.; Kothapalli, S.R.; Vaithilingam, S.; Oralkan, Ö.; Kamaya, A.; Wygant, I.O.; Zhuang, X.; Gambhir, S.S.; Jeffrey, R.B.; Khuri-Yakub, B.T. 3-D deep penetration photoacoustic imaging with a 2-D CMUT array. In Proceedings of the 2010 IEEE International Ultrasonics Symposium, San Diego, CA, USA, 11–14 October 2010.
126. Chen, J.; Wang, M.; Cheng, J.-C.; Wang, Y.-H.; Li, P.-C.; Cheng, X. A photoacoustic imager with light illumination through an infrared-transparent silicon CMUT array. *IEEE Trans. Ultrason. Ferroelectr. Freq. Control* **2012**, *59*, 766–775. [CrossRef] [PubMed]
127. Zhang, X.; Wu, X.; Adelegan, O.J.; Yamaner, F.Y.; Oralkan, Ö. Backward-mode photoacoustic imaging using illumination through a CMUT with improved transparency. *IEEE Trans. Ultrason. Ferroelectr. Freq. Control* **2017**, *65*, 85–94. [CrossRef]
128. Li, Z.; Ilkhechi, A.K.; Zemp, R. Transparent capacitive micromachined ultrasonic transducers (CMUTs) for photoacoustic applications. *Opt. Express* **2019**, *27*, 13204–13218. [CrossRef]
129. Matsumoto, Y.; Asao, Y.; Sekiguchi, H.; Yoshikawa, A.; Ishii, T.; Nagae, K.-I.; Kobayashi, S.; Tsuge, I.; Saito, S.; Takada, M. Visualising peripheral arterioles and venules through high-resolution and large-area photoacoustic imaging. *Sci. Rep.* **2018**, *8*, 1–11. [CrossRef]
130. Qiu, Y.; Gigliotti, J.V.; Wallace, M.; Griggio, F.; Demore, C.E.; Cochran, S.; Trolier-McKinstry, S. Piezoelectric micromachined ultrasound transducer (PMUT) arrays for integrated sensing, actuation and imaging. *Sensors* **2015**, *15*, 8020–8041. [CrossRef]
131. Griggio, F.; Demore, C.E.; Kim, H.; Gigliotti, J.; Qiu, Y.; Jackson, T.N.; Choi, K.; Tutwiler, R.L.; Cochran, S.; Trolier-McKinstry, S. Micromachined diaphragm transducers for miniaturised ultrasound arrays. In Proceedings of the 2012 IEEE International Ultrasonics Symposium, Dresden, Germany, 7–10 October 2012.
132. Akasheh, F.; Myers, T.; Fraser, J.D.; Bose, S.; Bandyopadhyay, A. Development of piezoelectric micromachined ultrasonic transducers. *Sens. Actuators A Phys.* **2004**, *111*, 275–287. [CrossRef]
133. Hajati, A.; Latev, D.; Gardner, D.; Hajati, A.; Imai, D.; Torrey, M.; Schoeppler, M. Three-dimensional micro electromechanical system piezoelectric ultrasound transducer. *Appl. Phys. Lett.* **2012**, *101*, 253101. [CrossRef]
134. Dangi, A.; Cheng, C.; Agrawal, S.; Tiwari, S.; Datta, G.R.; Benoit, R.; Pratap, R.; Trolier-McKinstry, S.; Kothapalli, S.-R. A photoacoustic imaging device using Piezoelectric Micromachined Ultrasound Transducers (PMUTs). *IEEE Trans. Ultrason. Ferroelectr. Freq. Control* **2020**, *67*, 801–809. [CrossRef] [PubMed]
135. Dangi, A.; Agrawal, S.; Tiwari, S.; Jadhav, S.; Cheng, C.; Trolier-McKinstry, S.; Pratap, R.; Kothapalli, S.-R. Evaluation of high frequency piezoelectric micromachined ultrasound transducers for photoacoustic imaging. In Proceedings of the 2018 IEEE SENSORS, New Delhi, India, 28–31 October 2018.
136. Dausch, D.E.; Gilchrist, K.H.; Carlson, J.B.; Hall, S.D.; Castellucci, J.B.; von Ramm, O.T. In vivo real-time 3-D intracardiac echo using PMUT arrays. *IEEE Trans. Ultrason. Ferroelectr. Freq. Control* **2014**, *61*, 1754–1764. [CrossRef] [PubMed]
137. Jung, G.; Tekes, C.; Rashid, M.W.; Carpenter, T.M.; Cowell, D.; Freear, S.; Degertekin, F.L.; Ghovanloo, M. A reduced-wire ICE catheter ASIC with Tx beamforming and Rx time-division multiplexing. *IEEE Trans. Biomed. Circuits Syst.* **2018**, *12*, 1246–1255. [CrossRef] [PubMed]
138. Fisher, R.; Thomenius, K.; Wodnicki, R.; Thomas, R.; Cogan, S.; Hazard, C.; Lee, W.; Mills, D.; Khuri-Yakub, B.; Ergun, A. Reconfigurable arrays for portable ultrasound. In Proceedings of the IEEE Ultrasonics Symposium, Rotterdam, The Netherlands, 18–21 September 2005.
139. Carpenter, T.M.; Rashid, M.W.; Ghovanloo, M.; Cowell, D.M.; Freear, S.; Degertekin, F.L. Direct digital demultiplexing of analog TDM signals for cable reduction in ultrasound imaging catheters. *IEEE Trans. Ultrason. Ferroelectr. Freq. Control* **2016**, *63*, 1078–1085. [CrossRef]

140. Brunner, E. *Ultrasound System Considerations and their Impact on Front-End Components*; Analog Devices: Norwood, MA, USA, 2002.
141. Gabrielli, A. Fast readout architectures for large arrays of digital pixels: Examples and applications. *Sci. World J.* **2014**, *6*, 523429. [CrossRef]
142. Freeman, S. TU-B-220-02: Microbeamforming for large-aperture ultrasound transducers. *Med. Phys.* **2011**, *38*, 3750–3751. [CrossRef]
143. Shabanimotlagh, M.; Raghunathan, S.; Bera, D.; Chen, Z.; Chen, C.; Daeichin, V.; Pertijs, M.; Bosch, J.G.; de Jong, N.; Verweij, M. Acoustic characterisation of a 32 × 32 element pzt-on-asic matrix transducer for 3D transesophageal echocardiography. In Proceedings of the 6th Dutch Bio-Medical Engineering Conference, Egmond aan Zee, The Netherlands, 26–27 January 2017.
144. Broschart, B. *Why Do Ultrasound Probes Cost So Much?* Conquest Imaging: Stockton, CA, USA, 2019.
145. Khairalseed, M.; Hoyt, K. Integration of a CMUT linear array for wideband H-scan ultrasound imaging. In Proceedings of the 2019 IEEE International Ultrasonics Symposium (IUS), Glasgow, Scotland, 6–9 October 2019.
146. Zheng, Z.; Na, S.; Albert, I.; Chen, H.; Li, Z.; Wong, L.L.; Sun, Z.; Yao, Y.; Liu, P.; Yeow, J.T. Development of a novel CMUT-based concentric dual-element ultrasonic transducer: Design, fabrication, and characterization. *J. Microelectromech. Syst.* **2018**, *27*, 538–546. [CrossRef]
147. Chan, J.; Zheng, Z.; Bell, K.; Le, M.; Reza, P.H.; Yeow, J.T. Photoacoustic imaging with capacitive micromachined ultrasound transducers: Principles and developments. *Sensors* **2019**, *19*, 3617. [CrossRef]
148. Kim, M.; Kim, J.; Cao, W. Aspect ratio dependence of electromechanical coupling coefficient of piezoelectric resonators. *Appl. Phys. Lett.* **2005**, *87*, 132901. [CrossRef]
149. Lindsey, B.D.; Kim, J.; Dayton, P.A.; Jiang, X. Dual-frequency piezoelectric endoscopic transducer for imaging vascular invasion in pancreatic cancer. *IEEE Trans. Ultrason. Ferroelectr. Freq. Control* **2017**, *64*, 1078–1086. [CrossRef]
150. Wang, X.; Pang, Y.; Ku, G.; Xie, X.; Stoica, G.; Wang, L.V. Noninvasive laser-induced photoacoustic tomography for structural and functional in vivo imaging of the brain. *Nat. Biotechnol.* **2003**, *21*, 803–806. [CrossRef] [PubMed]
151. Cheng, X.; Chen, J.; Li, C. A miniature capacitive micromachined ultrasonic transducer array for minimally invasive photoacoustic imaging. *J. Microelectromech. Syst.* **2010**, *19*, 1002–1011. [CrossRef]
152. Deferrari, H.A.; Darby, R.A.; Andrews, F.A. Vibrational displacement and mode-shape measurement by a laser interferometer. *J. Acoust. Soc. Am.* **1967**, *42*, 982–990. [CrossRef]
153. Deferrari, H.A.; Andrews, F.A. Laser interferometric technique for measuring small-order vibration displacements. *J. Acoust. Soc. Am.* **1966**, *39*, 979–980. [CrossRef]
154. Paltauf, G.; Nuster, R.; Haltmeier, M.; Burgholzer, P. Photoacoustic tomography using a Mach-Zehnder interferometer as an acoustic line detector. *Appl. Opt.* **2007**, *46*, 3352–3358. [CrossRef] [PubMed]
155. Bauer-Marschallinger, J.; Felbermayer, K.; Hochreiner, A.; Grün, H.; Paltauf, G.; Burgholzer, P.; Berer, T. Low-cost parallelization of optical fiber based detectors for photoacoustic imaging. In Proceedings of the Photons Plus Ultrasound: Imaging and Sensing 2013, San Francisco, CA, USA, 3–5 February 2013.
156. Thomson, J.K.; Wickramasinghe, H.K.; Ash, E.A. A Fabry-Perot acoustic surface vibration detector-application to acoustic holography. *J. Phys. D Appl. Phys.* **1973**, *6*, 677. [CrossRef]
157. Wickramasinghe, H.K. High frequency acoustic holography in solids. In *Acoustical Holography*; Springer: Berlin/Heidelberg, Germany, 1974; pp. 121–132.
158. Bucaro, J.A.; Dardy, H.D.; Carome, E.F. Fiber-optic hydrophone. *J. Acoust. Soc. Am.* **1977**, *62*, 1302–1304. [CrossRef]
159. Shajenko, P.; Flatley, J.P.; Moffett, M.B. On fiber-optic hydrophone sensitivity. *J. Acoust. Soc. Am.* **1978**, *64*, 1286–1288. [CrossRef]
160. Layton, M.R.; Bucaro, J.A. Optical fiber acoustic sensor utilizing mode-mode interference. *Appl. Opt.* **1979**, *18*, 666–670. [CrossRef] [PubMed]
161. Nuster, R.; Holotta, M.; Kremser, C.; Grossauer, H.; Burgholzer, P.; Paltauf, G. Photoacoustic microtomography using optical interferometric detection. *J. Biomed. Opt.* **2010**, *15*, 021307. [CrossRef] [PubMed]
162. Zhang, E.; Laufer, J.; Beard, P. Backward-mode multiwavelength photoacoustic scanner using a planar Fabry-Perot polymer film ultrasound sensor for high-resolution three-dimensional imaging of biological tissues. *Appl. Opt.* **2008**, *47*, 561–577. [CrossRef] [PubMed]

163. Ashkenazi, S.; Hou, Y.; Buma, T.; O'Donnell, M. Optoacoustic imaging using thin polymer etalon. *Appl. Phys. Lett.* **2005**, *86*, 134102. [CrossRef]
164. Maxwell, A.; Huang, S.-W.; Ling, T.; Kim, J.-S.; Ashkenazi, S.; Guo, L.J. Polymer microring resonators for high-frequency ultrasound detection and imaging. *IEEE J. Sel. Top. Quantum Electron.* **2008**, *14*, 191–197. [CrossRef]
165. Chao, C.-Y.; Ashkenazi, S.; Huang, S.-W.; O'Donnell, M.; Guo, L.J. High-frequency ultrasound sensors using polymer microring resonators. *IEEE Trans. Ultrason. Ferroelectr. Freq. Control* **2007**, *54*, 957–965. [CrossRef]
166. Leinders, S.M.; Westerveld, W.J.; Pozo, J.; van Neer, P.; Urbach, H.; de Jong, N.; Verweij, M.D. Design and characterization of a sensitive optical micro-machined ultrasound transducer. *J. Acoust. Soc. Am.* **2015**, *137*, 2426. [CrossRef]
167. Wu, Q.; Okabe, Y. High-sensitivity ultrasonic phase-shifted fiber Bragg grating balanced sensing system. *Opt. Express* **2012**, *20*, 28353–28362. [CrossRef]
168. Wang, D.H.; Jia, P.G.; Ma, Z.; Xie, L.; Liang, Q. Tip-sensitive fibre-optic Bragg grating ultrasonic hydrophone for measuring high-intensity focused ultrasound fields. *Electron. Lett.* **2014**, *50*, 649–650. [CrossRef]
169. Shnaiderman, R.; Wissmeyer, G.; Seeger, M.; Soliman, D.; Estrada, H.; Razansky, D.; Rosenthal, A.; Ntziachristos, V. Fiber interferometer for hybrid optical and optoacoustic intravital microscopy. *Optica* **2017**, *4*, 1180–1187. [CrossRef]
170. Parsons, J.E.; Cain, C.A.; Fowlkes, J.B. Cost-effective assembly of a basic fiber-optic hydrophone for measurement of high-amplitude therapeutic ultrasound fields. *J. Acoust. Soc. Am.* **2006**, *119*, 1432–1440. [CrossRef] [PubMed]
171. Hajireza, P.; Shi, W.; Bell, K.; Paproski, R.J.; Zemp, R.J. Non-interferometric photoacoustic remote sensing microscopy. *Light Sci. Appl.* **2017**, *6*, e16278. [CrossRef]
172. Maswadi, S.M.; Ibey, B.L.; Roth, C.C.; Tsyboulski, D.A.; Beier, H.T.; Glickman, R.D.; Oraevsky, A.A. All-optical optoacoustic microscopy based on probe beam deflection technique. *Photoacoustics* **2016**, *4*, 91–101. [CrossRef] [PubMed]
173. Barnes, R.A.; Maswadi, S.; Glickman, R.; Shadaram, M. Probe beam deflection technique as acoustic emission directionality sensor with photoacoustic emission source. *Appl. Opt.* **2014**, *53*, 511. [CrossRef] [PubMed]
174. Zanelli, C.I.; Howard, S.M. Schlieren metrology for high frequency medical ultrasound. *Ultrasonics* **2006**, *44*, e105–e107. [CrossRef]
175. Beard, P. Biomedical photoacoustic imaging. *Interface Focus* **2011**, *1*, 602–631. [CrossRef]
176. Nuster, R.; Schmitner, N.; Wurzinger, G.; Gratt, S.; Salvenmoser, W.; Meyer, D.; Paltauf, G. Hybrid photoacoustic and ultrasound section imaging with optical ultrasound detection. *J. Biophotonics* **2013**, *6*, 549–559. [CrossRef]
177. Preisser, S.; Rohringer, W.; Liu, M.; Kollmann, C.; Zotter, S.; Fischer, B.; Drexler, W. All-optical highly sensitive akinetic sensor for ultrasound detection and photoacoustic imaging. *Biomed. Opt. Express* **2016**, *7*, 4171–4186. [CrossRef]
178. Wurzinger, G.; Nuster, R.; Paltauf, G. Combined photoacoustic, pulse-echo laser ultrasound, and speed-of-sound imaging using integrating optical detection. *J. Biomed. Opt.* **2016**, *21*, 86010. [CrossRef]
179. Rousseau, G.; Gauthier, B.; Blouin, A.; Monchalin, J.P. Non-contact biomedical photoacoustic and ultrasound imaging. *J. Biomed. Opt.* **2012**, *17*, 61217. [CrossRef]
180. Nuster, R.; Gratt, S.; Passler, K.; Grün, H.; Berer, T.; Burgholzer, P.; Paltauf, G. Comparison of optical and piezoelectric integrating line detectors. In Proceedings of the Photons Plus Ultrasound: Imaging and Sensing 2009, San Jose, CA, USA, 25–28 January 2009.
181. Hamilton, J.D.; O'Donnell, M. High frequency ultrasound imaging with optical arrays. *IEEE Trans. Ultrason. Ferroelectr. Freq. Control* **1998**, *45*, 216–235. [CrossRef]
182. Beard, P.C.; Perennes, F.; Mills, T.N. Transduction mechanisms of the Fabry-Perot polymer film sensing concept for wideband ultrasound detection. *IEEE Trans. Ultrason. Ferroelectr. Freq. Control* **1999**, *46*, 1575–1582. [CrossRef]
183. Li, H.; Dong, B.; Zhang, Z.; Zhang, H.F.; Sun, C. A transparent broadband ultrasonic detector based on an optical micro-ring resonator for photoacoustic microscopy. *Sci. Rep.* **2014**, *4*, 4496. [CrossRef] [PubMed]
184. Rosenthal, A.; Razansky, D.; Ntziachristos, V. High-sensitivity compact ultrasonic detector based on a pi-phase-shifted fiber Bragg grating. *Opt. Lett.* **2011**, *36*, 1833–1835. [CrossRef] [PubMed]

185. Govindan, V.; Ashkenazi, S. Bragg waveguide ultrasound detectors. *IEEE Trans. Ultrason. Ferroelectr. Freq. Control* **2012**, *59*, 2304–2311. [CrossRef] [PubMed]
186. Zhu, X.; Zhu, X.; Huang, Z.; Wang, G.; Li, W.; Zou, D.; Li, C. Ultrasonic detection based on polarization-dependent optical reflection. *Opt. Lett.* **2017**, *42*, 439–441. [CrossRef]
187. Hariri, A.; Fatima, A.; Mohammadian, N.; Mahmoodkalayeh, S.; Ansari, M.A.; Bely, N.; Avanaki, M.R. Development of low-cost photoacoustic imaging systems using very low-energy pulsed laser diodes. *J. Biomed. Opt.* **2017**, *22*, 75001. [CrossRef]
188. Meimani, N.; Abani, N.; Gelovani, J.; Avanaki, M.R. A numerical analysis of a semi-dry coupling configuration in photoacoustic computed tomography for infant brain imaging. *Photoacoustics* **2017**, *7*, 27–35. [CrossRef]
189. Mohammadi, L.; Manwar, R.; Behnam, H.; Tavakkoli, J.; Avanaki, M.R.N. Skull's aberration modeling: Towards photoacoustic human brain imaging. In Proceedings of the Photons Plus Ultrasound: Imaging and Sensing 2019, San Francisco, CA, USA, 3–6 February 2019.
190. Mohammadi-Nejad, A.-R.; Mahmoudzadeh, M.; Hassanpour, M.S.; Wallois, F.; Muzik, O.; Papadelis, C.; Hansen, A.; Soltanian-Zadeh, H.; Gelovani, J.; Nasiriavanaki, M. Neonatal brain resting-state functional connectivity imaging modalities. *Photoacoustics* **2018**, *10*, 1–19. [CrossRef]
191. Xia, J.; Li, G.; Wang, L.; Nasiriavanaki, M.; Maslov, K.; Engelbach, J.A.; Garbow, J.R.; Wang, L.V. Wide-field two-dimensional multifocal optical-resolution photoacoustic-computed microscopy. *Opt. Lett.* **2013**, *38*, 5236–5239. [CrossRef]
192. Yao, J.; Xia, J.; Maslov, K.I.; Nasiriavanaki, M.; Tsytsarev, V.; Demchenko, A.V.; Wang, L.V. Noninvasive photoacoustic computed tomography of mouse brain metabolism in vivo. *NeuroImage* **2013**, *64*, 257–266. [CrossRef]
193. Kratkiewicz, K.; Manwar, R.; Rajabi-Estarabadi, A.; Fakhoury, J.; Meiliute, J.; Daveluy, S.; Mehregan, D.; Avanaki, K.M. Photoacoustic/ultrasound/optical coherence tomography evaluation of melanoma lesion and healthy skin in a swine model. *Sensors* **2019**, *19*, 2815. [CrossRef]
194. Kratkiewicz, K.; Manwar, R.; Zafar, M.; Mohsen Ranjbaran, S.; Mozaffarzadeh, M.; de Jong, N.; Ji, K.; Avanaki, K. Development of a stationary 3D photoacoustic imaging system using sparse single-element transducers: Phantom study. *Appl. Sci.* **2019**, *9*, 4505. [CrossRef]
195. Raj, B.; Rajendran, V.; Palanichamy, P. *Science and Technology of Ultrasonics*; Alpha Science Int'l Ltd.: Oxford, UK, 2004.
196. Zamora, I.; Ledesma, E.; Uranga, A.; Barniol, N. Miniaturized 0.13-µm CMOS front-end analog for AlN PMUT arrays. *Sensors* **2020**, *20*, 1205. [CrossRef] [PubMed]
197. Mouley, J. *Ultrasound Sensing Technologies for Medical, Industrial & Consumer Applications Report*; Yole Développement: Cornelius, NC, USA, 2018.
198. Allman, D.; Reiter, A.; Bell, M. Exploring the effects of transducer models when training convolutional neural networks to eliminate reflection artifacts in experimental photoacoustic images. In Proceedings of the Photons Plus Ultrasound: Imaging and Sensing 2018, San Francisco, CA, USA, 28 January–1 February 2018.
199. Singh, M.K.A.; Steenbergen, W. Photoacoustic-guided focused ultrasound (PAFUSion) for identifying reflection artifacts in photoacoustic imaging. *Photoacoustics* **2015**, *3*, 123–131. [CrossRef] [PubMed]
200. Li, R.; Phillips, E.; Wang, P.; Goergen, C.J.; Cheng, J.X. Label-free in vivo imaging of peripheral nerve by multispectral photoacoustic tomography. *J. Biophotonics* **2016**, *9*, 124–128. [CrossRef] [PubMed]
201. Sangha, G.S.; Phillips, E.H.; Goergen, C.J. In vivo photoacoustic lipid imaging in mice using the second near-infrared window. *Biomed. Opt. Express* **2017**, *8*, 736–742. [CrossRef]
202. Xu, Q.; Jalilian, E.; Fakhoury, J.W.; Manwar, R.; Michniak-Kohn, B.; Elkin, K.B.; Avanaki, K. Monitoring the topical delivery of ultrasmall gold nanoparticles using optical coherence tomography. *Ski. Res. Technol.* **2019**, *26*, 263–268. [CrossRef]
203. Oh, B.-H.; Kim, K.H.; Chung, K.-Y. Skin imaging using ultrasound imaging, optical coherence tomography, confocal microscopy, and two-photon microscopy in cutaneous oncology. *Front. Med.* **2019**, *6*, 274. [CrossRef]
204. Elson, D.S.; Li, R.; Dunsby, C.; Eckersley, R.; Tang, M.-X. Ultrasound-mediated optical tomography: A review of current methods. *Interface Focus* **2011**, *1*, 632–648. [CrossRef]
205. Fujimoto, J.G.; Pitris, C.; Boppart, S.A.; Brezinski, M.E. Optical coherence tomography: An emerging technology for biomedical imaging and optical biopsy. *Neoplasia* **2000**, *2*, 9–25. [CrossRef]

206. Shu, X.; Beckmann, L.; Wang, Y.; Rubinoff, I.; Lucy, K.; Ishikawa, H.; Wollstein, G.; Fawzi, A.A.; Schuman, J.S.; Kuranov, R.V. Designing visible-light optical coherence tomography towards clinics. *Quant. Imaging Med. Surg.* **2019**, *9*, 769. [CrossRef]
207. Jalilian, E.; Xu, Q.; Horton, L.; Fotouhi, A.; Reddy, S.; Manwar, R.; Daveluy, S.; Mehregan, D.; Gelovani, J.; Avanaki, K. Contrast-enhanced optical coherence tomography for melanoma detection: An in vitro study. *J. Biophotonics* **2020**, *13*, e201960097. [CrossRef] [PubMed]
208. Mahmoodkalayeh, S.; Zarei, M.; Ansari, M.A.; Kratkiewicz, K.; Ranjbaran, M.; Manwar, R.; Avanaki, K. Improving vascular imaging with co-planar mutually guided photoacoustic and diffuse optical tomography: A simulation study. *Biomed. Opt. Exp.* **2020**, *11*, 4333–4347. [CrossRef]

© 2020 by the authors. Licensee MDPI, Basel, Switzerland. This article is an open access article distributed under the terms and conditions of the Creative Commons Attribution (CC BY) license (http://creativecommons.org/licenses/by/4.0/).

micromachines

MDPI

Review

Recent Progress on Extended Wavelength and Split-Off Band Heterostructure Infrared Detectors

Hemendra Ghimire [1], P. V. V. Jayaweera [2], Divya Somvanshi [3], Yanfeng Lao [4] and A. G. Unil Perera [1,*]

1 Center for Nano-Optics (CeNo), Department of Physics and Astronomy, Georgia State University, Atlanta, GA 30033, USA; hghimire1@student.gsu.edu
2 SPD Laboratory, Inc., Hamamatsu 432-8011, Japan; virajjayaweera@gmail.com
3 Department of Electronics and Tele-Communication Engineering, Jadavpur University, Kolkata 700032, India; somvanshi.divya@gmail.com
4 Hisense Photonics, Inc., 5000 Hadley Road, South Plainfield, NJ 07080, USA; ylao95@gmail.com
* Correspondence: uperera@gsu.edu

Received: 21 April 2020; Accepted: 20 May 2020; Published: 28 May 2020

Abstract: The use of multilayer semiconductor heterojunction structures has shown promise in infrared detector applications. Several heterostructures with innovative compositional and architectural designs have been displayed on emerging infrared technologies. In this review, we aim to illustrate the principles of heterostructure detectors for infrared detection and explore the recent progress on the development of detectors with the split-off band and threshold wavelength extension mechanism. This review article includes an understanding of the compositional and the architectural design of split-off band detectors and to prepare a database of their performances for the wavelength extension mechanism. Preparing a unique database of the compositional or architectural design of structures, their performance, and penetrating the basics of infrared detection mechanisms can lead to significant improvements in the quality of research. The brief outlook of the fundamentals of the infrared detection technique with its appropriateness and limitations for better performance is also provided. The results of the long-term study presented in this review article would be of considerable assistance to those who are focused on the heterostructure infrared detector development.

Keywords: heterostructures; split-off band; wavelength extension; device performance

1. Introduction

Technological advancement in material growth, processing, and characterization setups, furnished with new concepts of device structures and functions, has been widely implemented for the development of new, low-disorder, and often highly engineered material combinations, such as heterostructures [1,2]. The using layers of structurally abrupt interfaces of dissimilar compounds artificially periodic structures can be obtained [3]. Many scientists have thus used refractive indexes, bandgap, effective masses and mobilities of charge carriers, and the electron energy spectrum advantages of semiconductor heterostructures with various material combinations, architectures, and doping densities for the futuristic scientific, technical and biomedical applications [4–10]. The examples include GaAs/AlGaAs heterostructures [11], which have been well-studied for its potential application in high-speed digital and optoelectronic devices [12] including diode lasers [13], light-emitting diodes [14], solar cells [15] and optical detectors [8,16–20]. The material advantage of GaAs/AlGaAs provides excellent uniformity and large arrays. It is because there is a close match between the lattice constant of AlGaAs and GaAs. Herein, AlGaAs has a lattice constant that varies linearly between that of AlAs, 5.661 angstroms, and that of GaAs is 5.653 angstroms, depending on the mole fraction of aluminum [21]. The alternations of the aluminum fraction in the AlGaAs layer modulate the band-gap [22] and hence, one can adjust the barrier height of the designed device to match the required photon energy. Similar to GaAs/AlGaAs [23], research has been carried out on the number of other semiconductor heterostructures having various, material combinations emitter/barrier architectures and doping densities for infrared (IR) detector development [6,24–28].

The heterojunction detectors offer wavelength flexibility [29–33] and multicolor capability in these regions. The wavelength range includes short-wave infrared (SWIR) of 1–3 μm [29], mid-wave infrared (MWIR) of 3–5 μm [30], long-wave infrared (LWIR) of 5–14 μm [33], very-long-wave infrared (VLWIR) of 14–30 μm [32], far-wave infrared (FWIR) of 30–100 μm or 3–10 THz [31] and so on. Among these wavelength ranges of IR radiations, 3–5 [34] and 8–14 μm [35] are considered the atmospheric windows and are suitable for IR applications.

In this review, we stepwise describe investigations concerning the physics and applications of split-off band semiconductors heterostructures infrared detectors for the threshold wavelength extension mechanism. The description systematically explains the architectural design of the device and assesses current experimental and theoretical understanding used to improve the performance.

1.1. Heterostructure IR Detectors

The architecture of heterostructures [36] mainly involves alternate thin layers of lattice-matched and band-gap tuned different compound semiconductors. The growth of sandwiches' layers follows appropriate fabrication so that unique (unlike from component semiconductors) electrical and optoelectrical properties of heterostructure can be achieved. Bandgap engineered compact heterostructure is then processed with appropriate conducting rings for characterization. The processing for the fabrication also includes photolithography [37], etching to open an optical window for incidence illumination, lift-off, and others. The typical architecture of the heterostructure detector is shown in Figure 1A. In such a compact structure, each layer of the emitter is sandwiched between two layers of barriers. A sharp interface between the junctions of each layer can be achieved by making it easier for tuning the energy gap between the emitter and barrier. The development of sophisticated controllable epitaxial growth methods, such as Molecular Beam Epitaxy (MBE) and Metal-Organic Chemical Vapor Deposition (MOCVD), has allowed for the fabrication of such ideal heterojunction structures [38–40].

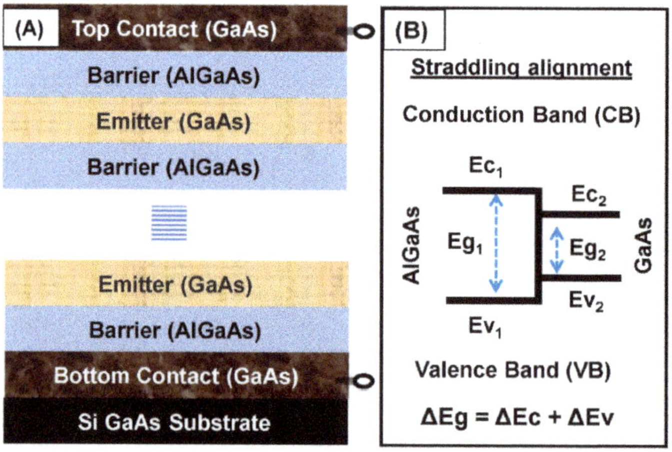

Figure 1. Typical architecture of a GaAs/AlGaAs heterostructure and the type depending on band alignments. (**A**) The combination of two dissimilar semiconductors in the heterostructure. The design is a p-GaAs/AlGaAs heterostructure and every layer of the emitter (p-type GaAs) is sandwiched between two layers of the barrier (AlGaAs). (**B**) Type-I heterostructure made of GaAs and AlGaAs. In such heterostructures, bandgap overlaps and $\Delta Eg = \Delta Ec + \Delta Ev$.

The emitter and barrier regions of heterostructure have different energy bands, and it will have an additional change in the presence of biasing [41]. In such compound heterostructures, quantum-mechanical effects such as potential discontinuity (band off-set) play a crucial role [42]. Based on the alignment of bands producing (step) discontinuity, heterostructures are classified into three types [36], type I, type II, and type III. Figure 1B shows the straddling alignment in a type-I heterostructure, where signs of the band are offset, as the two bands are opposite. In type-I heterostructures, the band gaps of one material entirely overlap with that of another, and potential discontinuities are as $\Delta Ec = Ec_1 - Ec_2$, $\Delta Ev = Ev_1 - Ev_2$ and ΔEg $(Eg_1 - Eg_2) = \Delta Ec + \Delta Ev$. Similarly, in a type-II heterostructure, $Ev_1 > Ev_2$ and ΔEc may or may not be larger than Eg_1. Moreover, Eg_2 is every so often smaller than Eg_1 in a type-II heterostructure. Type II staggered, (where $\Delta Ec < Eg_1$) and type II misaligned (where $\Delta Ec > Eg_1$) are two subclasses of type-II heterostructures. Type-III heterostructures are formed by combining semimetal with the inverted bands of semiconductors [43]. In this paper, we have mainly used GaAs/AlGaAs to cover type I heterostructures as shown in the above figure.

Along with technological development, there is a growing demand for advanced IR systems with better discrimination and identification. Group-IV, III-V, and II-VI semiconductor heterostructure-based [44,45] photodetectors have been studied extensively for the IR detection, from near infrared (NIR) to far infrared (FIR) region [46]. SiGe/Si heterojunctions were used to study internal photoemission LWIR detectors [33,47]. Low dimensional II-VI oxides semiconductor structures such as NiO/ZnO were also studied to understand their feasibility for light detection [48]. Similarly, studies of III-V semiconductor alloys such as InGaAs, InAsSb, InGaSb, HgZnTe, HgMnTe, GaAs, AlGaAs and their heterostructures are mainly focusing on the MWIR and LWIR detectors [49,50]. Mercury alloys led tin tellurides and selenides, and other alloy combinations were further explored for their potential as alternative material that can overcome the challenges of important semiconductor IR detector mercury cadmium telluride (HgCdTe or MCT) [27,50–52]. HgCdTe-based IR detectors are the most widely used for high-performance applications [50]. However, the progress has been impeded due to the fundamental material problems, such as high defect density related to growth due to the weak HgTe bond [53]. This is a serious technological problem for the mass production of HgCdTe large-sized Focal Plane Arrays (FPA), and to achieve high FPA pixel uniformity and yield.

On the other hand, Quantum Well IR Photodetectors (QWIPs), based on the GaAs/AlGaAs material system, are considered mostly for the LWIR spectral regime requiring high uniformity. The industrial infrastructure in III–V materials/device growth, processing, and packaging brought about by the utility of GaAs-based devices in the telecommunications industry gives QWIPs a potential advantage in producibility and cost [50]. However, QWIPs are not sensitive to normal incident light due to orthogonality between polarization vector of the incident photons and optical transition dipole moment [54]; therefore, light coupling structures are required for a 45° angle incidence of light which adds cost and complexity. Besides, a low operating temperature is typically required due to fundamental limitations associated with intersubband transitions [50]. Studies of Quantum Dot IR Photodetectors (QDIPs) are also making progress towards the development of IR detectors [50,55].

Because of the key pieces of evidence presented above, the studies show number of different structures and material combinations were used. The trend of exploration of new material is still increasing day by day. The superiority of heterojunction structures over the homojunction further allows for the estimation of new materials and engineering compositions in the coming days.

1.2. Performances of IR Detectors and Acceptable Figures of Merits

Studies have illustrated a wide range of accepted figures of merit for a meaningful comparison of the sensitivity performances of detectors [56]. These include responsivity, spectral response, specific detectivity, dark current, and the current gain. Among these performance matrices, responsivity measures the input-output gain (simply called gain) of the detector. Simply put, it is defined [57] as the ratio of the photocurrent (Ipc) and incoming optical power (Io), i.e., $R = Ipc/Io = \eta q \lambda / hv \approx \eta \lambda / 1.24$ (A/W), where $hv = hc/\lambda$ is the energy of the incident photon, h planks constant, c velocity of light, the q charge of an electron, λ is the wavelength of incident light and η is quantum efficiency. Studies have used spectral response [58] to describe the sensitivity (ability to convert light of various wavelengths to electricity) of the photodetectors as a function of photon frequency (or wavelength). The efficiency of detection is also simply expressed in terms of quantum efficiency in various studies [59,60]. Specific detectivity [56] is another important performance matrix, which is defined as the reciprocal of noise equivalent power (NEP) normalized pre-square root of the sensor's area (A) and frequency (f) bandwidth, i.e., $D^* = \frac{\sqrt{Af}}{NEP}$. Similarly, the current flowing through the detector in the absence of light, known as the dark current, is another important parameter to determine the performance of the detector [61]. The dark current characteristics are important to determine an optimum operating condition. Whilst the lowest possible dark current is desired for the operation of an IR photodetector, a certain amount of bias voltage must be applied to operate photoconductive photodetectors [61]. The study of dark current characteristics offers important insights into the device parameters, such as activation energy (Δ). Dark current in the semiconductor heterostructures is temperature dependent, which can be expressed [62,63] as:

$$I_{dark} = Ae\frac{\mu F}{\left[1+\left(\frac{\mu F}{v_{sat}}\right)^2\right]^{1/2}}2\left(\frac{m^* k_B T}{2\pi\hbar^2}\right)^{3/2}\exp\left(-\frac{E_{act}}{k_B T}\right) \quad (1)$$

where A is the electrically active area of the detector, e is the electronic charge, $E_{act} = \Delta - Af - E_f$ is the activation energy, μ is the mobility of the holes, v_{sat} is the saturation velocity, m^* is the effective mass, k_B is Boltzmann's constant, T is the temperature, $\hbar$ is the reduced Planck constant, α is a constant parameter that determines effective barrier lowering due to the applied field F, and E_f is the Fermi level.

1.3. Spin-Orbit Split-Off Band Heterostructures Infrared Detectors

The use of semiconductor heterostructures for the detection of electromagnetic (EM) spectrum blind to the human eye (UV, IR, and THz) [64] leads towards the exploration of structural concepts, their physics, theory, modeling, and experimental measurements. The study of intraband electronic transition within the valance band (heavy hole (HH), light hole (LH) and spin-orbit/split-off (SO) bands) is one of the primary characteristic features to understand the electronic process in semiconductor alloys [65,66]. Split-off band effects have been observed in the emission of GaAs metal-semiconductor field-effect transistors [67] and have enhanced the response of GaInAsP quantum wells [68]. This aspect of career transition is also implemented to develop heterostructure infrared detectors [69].

A detailed explanation of experimental SO response depending on hole transitions between LH, HH and SO bands has been presented by using GaAs/AlGaAs-based heterojunction interfacial workfunction internal photoemission (HEIWIP) detectors [70–72]. The active region of the basic HEIWIP detector consists of one or more periods, each consisting of a doped emitter and an undoped barrier layer. These multiple emitter/barrier layers are sandwiched between two highly doped contact layers. Depending on the doping required for ohmic contacts, the top contact may also serve as the top emitter layer. Herein, the work function (Δ) is given by $\Delta = \Delta_d + \Delta_x$, where Δ_d and Δ_x are the contributions from doping and the Al fraction, respectively [23]. As the Al fraction is reduced, Δ will be limited by Δ_d, which in turn is a homojunction detector [73,74]. The detection mechanism can be divided into three main processes: (i) the photoabsorption that generates excited carriers, (ii) the escape of the carriers, and (iii) a collection of the escaped carriers.

The schematic of different intervalence and intra-band optical transitions showing the infrared detection mechanism is shown in Figure 2 in terms of energy wave vector diagrams. Figure 2A–C shows the energy wavenumber (E-k) diagram for intrinsic, extrinsic, and quantum well infrared photodetectors. Figure 2D shows Split-off detector threshold mechanisms. Infrared photon excites holes from the light/heavy hole bands to the split-off band. (1) Indirect absorption followed by scattering and escape (threshold energy: $E_{ESO} - E_f$). (2) Direct absorption followed by scattering and escape (threshold energy: $E_{ESOf} - E_f$). (3) Indirect absorption followed by escape and some scattered (threshold energy: $E_{BSO} - E_f$). The band alignment during split-off band intra-valance transitions is shown in Figure 3F.

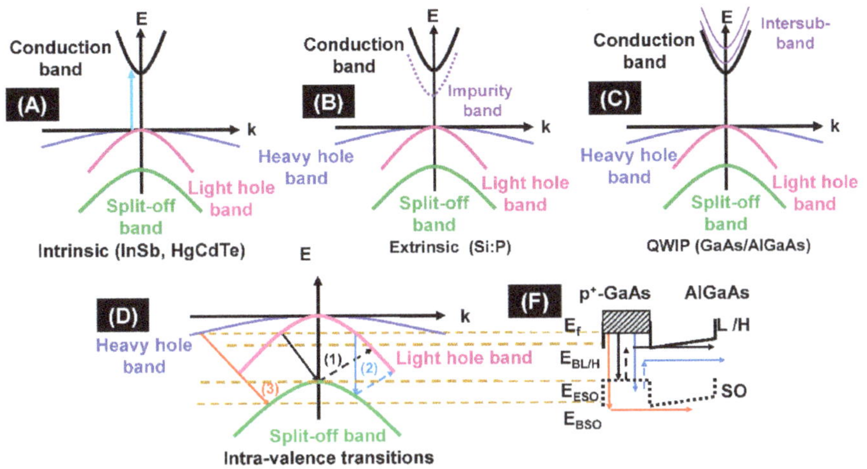

Figure 2. Schematic of different intervalence and intraband optical transitions showing an infrared (IR) detection mechanism. (**A–C**) Energy wavenumber (E-k) diagram for intrinsic, extrinsic, and quantum-well-IR photodetectors, respectively. (**D**) Split-off detector threshold mechanisms, where the IR photon excites holes from the light/heavy hole bands to the split-off band. (1) Indirect absorption followed by scattering and escape (threshold energy: $E_{ESO} - E_f$). (2) Direct absorption followed by scattering and escape (threshold energy: $E_{ESO} - E_f$). (3) Indirect absorption followed by escape and some scattered (threshold energy: $E_{BSO} - E_f$). (**F**) Band alignment using during split-off band intra-valance transitions.

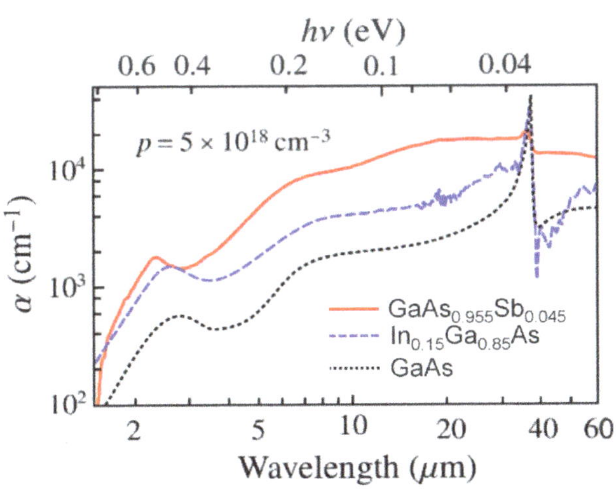

Figure 3. Absorption coefficient of p-type (5×10^{18} cm^{-3}) $GaAs_{0.955}Sb_{0.045}$, $In_{0.15}Ga_{0.85}As$ and GaAs as discussed [69]. The p-type $GaAs_{0.955}Sb_{0.045}$ shows the highest absorption coefficient by comparing with the other two materials.

The spectral response of the heterojunction detector is primarily determined by the absorbing properties of the emitter, which is, in turn, determined by the electronic structure of the valence bands (VBs) and the concentration of holes. Utilizing semiconductor heterojunctions for the heterojunction detectors is subject to limitations, such as low absorption coefficient in the 3–5 μm range for p-type GaAs. The p-type semiconductor absorbs extra photons owing to optical transitions among three VBs, i.e., the LH, HH, and SO bands, contributing to an absorption band spanning primarily from 1 to 10 μm for p-type GaAs, as shown in Figure 3. The p-type semiconductors including InGaAs, GaAsSb, and GaAs indicate that the shifting of the absorption peak (between 1 and 5 μm) can be associated with the SO–HH transition. The intra-band free-carrier absorption is typically proportional to λ^p, which becomes dominant in the very-long-wavelength range, where λ is the wavelength, and p is an exponent. p can be predicted by the Drude theory, and usually equals 1.2, 2.5, and 3.5 correspondings to scatterings by acoustic phonons, polar optical phonons, and ionized impurities, respectively. By comparing with p-type GaAs, p-type GaAsSb shows a higher absorption, which benefits a higher absorption efficiency and hence the detector performance.

Jayaweera et al. studied an uncooled infrared detector for 3–5 μm and beyond [75] by using the concept of a split-off band infrared detection mechanism [70,71]. In their study, they have experimentally demonstrated uncooled infrared detection using intra-valence bands using a set of three p-GaAs/AlGaAs heterostructure. The focus of their study is to demonstrate an uncooled infrared detection using intra-valence bands. The uncooled detection of infrared radiation is important in a wide range of applications in the civilian, industrial, medical, astronomy, and military sectors. The calculated D^* value in this study is 6.8×10^5 Jones at a temperature of 300 K and an SO band offset of 0.31 eV, while, for those with an SO band offset of 0.155 and 0.207 eV, D^* values are 2.1×10^6 and 1.8×10^6 at temperatures of 140 and 190 K, respectively. It is noted that the studies on the high operating temperature that split off the transition in such a heterostructure [70–72] further support the infrared detection mechanism. These studies have introduced intra-valence band transitions, such as LH to heavy-hole HH transitions or HH transitions to SO band in detectors to overcome the selection rule limitations. A band diagram (E-k) of the emitter region of an S-O band detector illustrating the detection mechanism is based on the carrier transitions in the three valence bands is also demonstrated.

In the study by Lao and et al. [76], the development of structure with multi-spectral detection probability is presented. The hole transition from the HH to the LH band of p-GaAs/AlGaAs detectors show a spectral response up to 16.5 m, operating up to a temperature of 330 K where the LH-HH response is superimposed on the free-carrier response. Similarly, in another study by Matsik, and et al. [77] device modeling study of S-O band IR detector was carried out to find optimized conditions for its performance improvement. The study suggested two important device architecture modifications in order to improve device performance. One of the suggestions was to include an offset (δE) between the energy barriers so that the low energy barrier towards the collector side would enhance the collection. The other suggestion was to include a graded barrier on the injector side so that the carrier trapping would be reduced and the injection of the carriers to the absorber/emitter would be improved.

Altogether, the peak response due to intra-valance band transitions [70,71,78] is believed to originate from a build-up of a quasi-equilibrium Fermi level, at a fixed level, irrespective of the variation of the device parameter. Thus, the SO band was found to be the most probable energy level to build-up a quasi-equilibrium Fermi level as a consequence of the hot-phonon bottleneck effect. The study of the dark current characteristics of these IR photodetectors confirmed no compromise in the dark current due to the presence of the extended-wavelength mechanism of the photoresponse. Furthermore, the quasi-equilibrium will be built-up not only in the emitter but also in the bottom and top contact layers as long as there is photoexcitation to the SO band. The net flow of the photo-excited carriers will determine the spectral photoresponse [62,79].

In summary, HEIWIP detectors [80], based on intra-valence band transitions LH to HH transitions or HH transitions to SO band were introduced to overcome the selection rule limitations. The SO band infrared detector is a newly developed, emerging device based on the p-doped GaAs/AlGaAs system [71], which utilizes hole transition in the HH/LH band to the SO band as the detection mechanism [71]. The SO band detectors have shown promising results to be developed as an uncooled IR detector [75].

1.4. *Effect of a Current Blocking Barrier on Heterojunction Infrared Detector*

Studies have shown that incorporating current blocking architectures into detector designs alters their performances. As part of their studies, Wang and et al. [81], Rotella and et al. [82], Wang and et al. [83], Lin and et al. [84], Pal and et al. [85] and Nevou and et al. [86] have presented the use of AlGaAs current blocking layers in quantum dot IR photodetectors (QDIPs) to enhance the performance of detectors. Similarly, Stiff et al. [87], Tang [88], and Chakrabarti [89] have presented the use of a similar concept to achieve higher operating temperatures. In another study by Nguyen et al. [90], hole blocking layers have been implemented in type-II InAs/GaSb superlattice infrared photodetectors. The electron blocking and hole blocking unipolar barriers in complementary barrier infrared detectors [91] and p-type-intrinsic-n-type photodiodes [92] were also studied. The most important goal in these architectures is to increase the performance by lowering the dark current with a relatively small compromise in the photocurrent.

Chauhan et al. have reported the performance of a p-GaAs/AlxGa1xAs heterojunction photovoltaic infrared detector, with graded barriers, operating in the 2–6 μm wavelength range [16]. They found that the incorporation of the current blocking barrier in heterostructure architecture leads to achieving a significant improvement in the specific detectivity (D^*). It further reduces the dark current under photoconductive operation and increases the resistance-area product (R_0A, R_0 is the resistance at 0 V and A is the electrical area) under photovoltaic operation. In blocking barrier studies, measurements across the top and bottom contacts include the current blocking barrier, whilst the middle and bottom contacts measure the same sample (mesa) without the current blocking barrier, as shown in the schematic diagram of the valence band of the heterostructure in Figure 4. Herein, the implementation of a current blocking barrier increases the specific detectivity (D^*) under dark conditions by two orders of magnitude to 1.9×10^{11} Jones, at 77 K. Furthermore, at zero bias, the resistance-area product (R_0A) attains a five orders enhancement due to the current blocking barrier, with the responsivity reduced by only a factor of 1.5.

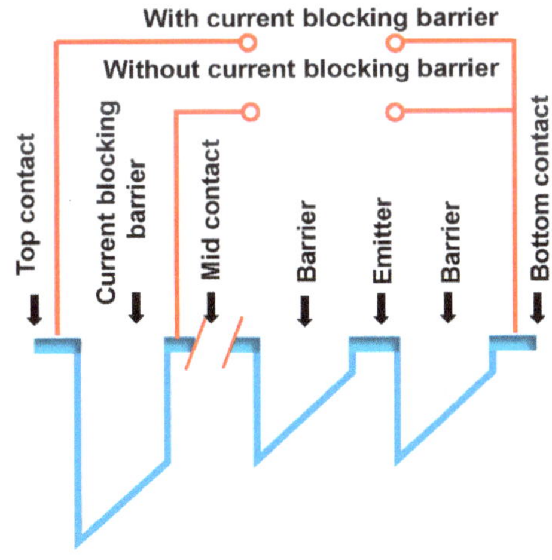

Figure 4. Schematic of the valence band alignment of the heterostructure under equilibrium showing the connections with and without the current blocking barrier.

1.5. *Threshold Wavelength Extension Mechanism*

The study of the S-O band detector devices with the barrier offset (δE) and a graded barrier showed an unprecedented result in terms of the spectral range of the photoresponse: a photoresponse that is far beyond the spectral limit. In general, the standard expected threshold wavelength (λ_t) of a photodetector is governed by the spectral rule $\Delta = 1.24/\lambda_t$, where Δ controls the dark current. The extended wavelength IR photodetectors are a novel class of photodetectors showing spectral photoresponse far beyond the conventional limit of $\Delta = hc/\lambda_t$. In extended wavelength detectors, the effective response threshold wavelength (λ_{eff}) is governed by a $\Delta' = 1.24/\lambda_{eff}$, with an effective activation energy $\Delta' \ll \Delta$. The valance band diagram used to demonstrate the extended wavelength photoresponse utilized an offset (δE) between the energy barriers in the heterostructure is shown in Figure 5B. A reference photodetector without the offset and cannot show the extended wavelength mechanism is also shown in Figure 5A.

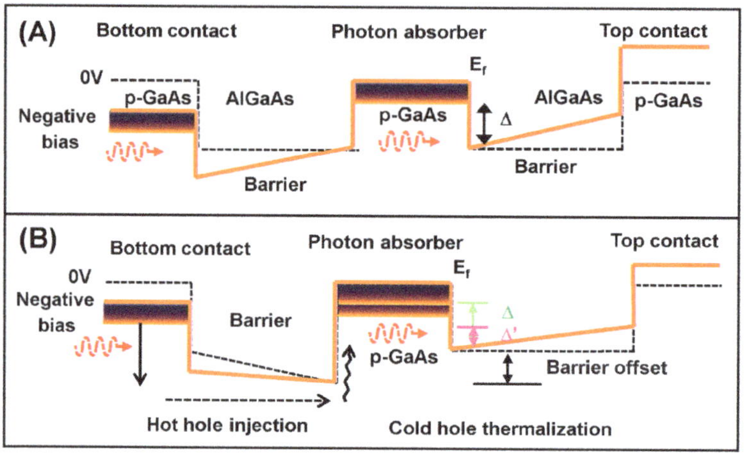

Figure 5. Conventional and wavelength extended photodetector. (**A**) Conventional photodetector. The threshold wavelength (λ_c) of detection is determined by Δ, where $\lambda_c = hc/\Delta$. (**B**) Wavelength extended photodetector showing barrier offset contributing to wavelength-extended photoresponse.

The mechanism responsible for the extension of threshold wavelength in heterostructure detectors has been analyzed by Somvanshi et al. [93], which is based on the hot carrier effects in the semiconductor heterostructures [94–96]. The hot carrier effect is principally governed by the carrier–carrier and carrier–phonon scattering processes and has been widely studied in the past [96–98]. In general, hot carriers are created in energy states above the band edge, interact with lattice vibrations, and cold carries through carrier–carrier interactions. This leads to quasi-equilibrium Fermi distributions at a temperature much higher than the lattice temperature [97,98]. It is observed that, upon the incidence of light from an external optical source, hot holes have created in the bottom contact, photon absorber, and top contact of heterostructure detectors, as shown in Figure 5. In heterostructure detectors with band offset (Figure 5B), a net flow of hot holes has observed towards the top contact, owing to the difference in barrier heights, ΔEv. However, no such flow of carriers observed in heterostructure detectors without band offset, i.e., conventional detectors (Figure 5A). The dynamics of hot hols and cold holes interaction in the p-GaAs photon absorber can be explained based on hot carrier effects. Upon the interaction of carriers in the photon absorber, the exchange of energy takes place through hole–hole and hole–phonon scattering that leads to the formation of a quasi-Fermi distribution (E_{quasiF}) at a hot hole temperature (T_H) that is significantly greater than the lattice temperature (T_L). The distribution of hot holes at this quasi-Fermi level will lead to an escape of hot holes across the collector barrier when a long-wavelength photon is absorbed and the extension in threshold wavelength is observed.

Chauhan et al. have studied the photoresponse of extended wavelength IR photodetectors [99] under different bias conditions. It is observed that, with the increase in bias voltages, the photoresponse becomes stronger; however, the spectral threshold remains relatively constant. The dark current and photo-response characteristics of extended wavelength infrared photodetectors [62] studied by Chauhan and et al. show that the measured dark current of extended wavelength detectors was found to agree well with fits obtained from a 3D carrier drift model using the designed value of Δ. In contrast, the spectral photoresponse showed extended wavelength thresholds corresponding to Δ′. Since the dark current in the extended-wavelength IR photodetectors is still limited by Δ as in the conventional photodetectors, the extended wavelength mechanism offers a new avenue for the design and development of the IR photodetectors.

To understand the role of energy offset for threshold wavelength extension, a few heterostructure semiconductors were comparatively analyzed using references [100–102]. Schematic diagrams of the valence band alignment of the detector designs under equilibrium are shown in Figure 6. Detectors without barrier offset, LH1002 is shown in Figure 6A. SP1001 (Figure 6B) consists of a p-GaAs emitter (80 nm), an 80 nm $Al_{0.75}Ga_{0.25}As$ barrier at the bottom, and a 400 nm $Al_{0.57}Ga_{0.43}As$ barrier at the top. These layers are sandwiched between highly doped p-GaAs top and bottom contact layers. SP1001 has an energy offset (δE) of ~0.10 eV between the barriers. In SP1007, 15SP3, GSU17I, GSU17II, and GSU17III, as shown in Figure 6C, the bottom AlxGa1-xAs barrier (80 nm) has a graded potential profile obtained by increasing the Al mole fraction from X_1 at the bottom of this layer to X_2 at the top. The top AlxGa1-xAs barrier (400 nm) has a constant barrier potential profile with $X_3 = X_4$. The emitters (80 nm) are thick enough to have a bulk-like distribution of energy states. Changing the X_3 causes the δE variation and changing X_1 causes the gradient variation, as summarized in Table 1.

In summary, these studies and the devices LH1002, SP1001, SP1007, 15SP3, GSU17I, GSU17II, and GSU17III were grown on semi-insulating GaAs substrate by molecular beam epitaxy. Each heterostructure device consists of an AlxGa1-xAs barrier at the bottom, followed by a p-GaAs emitter, and then another AlxGa1-xAs barrier at the top. The bottom AlxGa1-xAs barrier is graded by increasing the Al mole fraction from a lower value x1 at the bottom of this layer to higher value X_2 at the top, except for SP1001 with $X_1 = X_2$ to form a constant barrier. These emitter/barrier layers are sandwiched between the bottom and top contact layers of p-type doped GaAs. The emitter and the top and bottom contact layers are degenerately p-type doped at 1×10^{19} cm^{-3}, whilst the AlxGa1-xAs barriers are undoped. The thickness of the p-GaAs emitter in all devices is sufficiently large to have a bulk-like distribution of energy states. 15SP3, GSU17I, and GSU17II constitute another set with varying values of the gradient ~20.6, 28.9, and 37.1 kV/cm, given by $(\Delta Ev(x_2) - \Delta Ev(x_1))/w_1$ for $x_1 = 0.45$, 0.33, and 0.21, respectively. The gradients are the only differences in these three devices. It should be noted that the device 15SP3 is common in both sets with an δE ~0.19 eV and gradient 20.6 kV/cm. The energy values (ΔTDIPS) corresponding to the determined wavelength threshold are also shown in Table 1. Herein, nearly doubling the barrier gradient from 20.6 to 37.1 kV/cm did not show a significant change in the wavelength threshold. These results confirm that the extended wavelength mechanism originating from the quasi-equilibrium Fermi level at a fixed energy level.

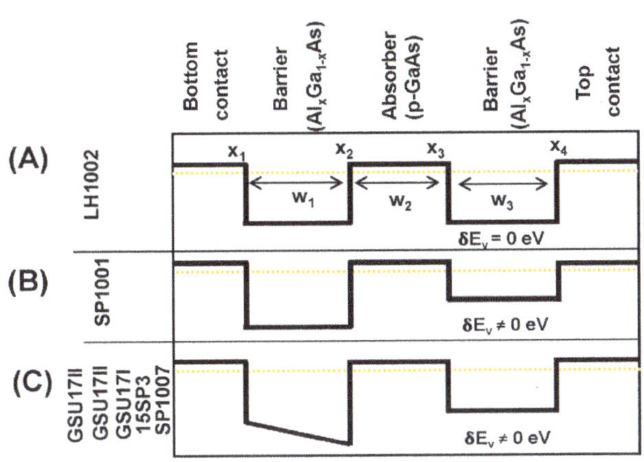

Figure 6. Schematic diagrams of the valence band alignment of the detector designs under equilibrium. (**A**) Detectors without barrier offset LH1002. (**B**) SP1001 consists of an emitter at the bottom, the barrier at the top, with an energy offset (δEv) of ~0.10 eV between the barriers. (**C**) In SP1007, 15SP3, GSU17I, GSU17II, and GSU17III have graded bottom by tuning the Al mole fraction and barrier offset δEv not equal to zero.

Table 1. Summary of results from the temperature-dependent internal photoemission spectroscopy (TDIPS) fitting method with corresponding device parameters (threshold wavelength is also shown).

Device	Δ (eV)	δE (eV)	A_l Fraction			Thickness (nm)			Gradient (kV/cm)	Δ_{TDIPS} (eV)	λt (μm)
			x_1	x_2	$x_3 = x_4$	We	W_1	W_2	$[\Delta Ev(x_2) - \Delta Ev(x_1)]/w_1$		
LH1002	0.3	0	0.57	0.57	0.57	20	20	60	0	0.2781 ± 0.0006	4.50 ± 0.01
SP1001	0.4	0.1	0.75	0.75	0.57	80	80	400	0	0.0248 ± 0.0001	50.0 ± 0.3
SP1007	0.4	0.1	0.45	0.75	0.57	80	80	400	20.6	0.0223 ± 0.000	56.0 ± 0.5
15SP3	0.4	0.19	0.45	0.75	0.39	80	80	400	20.6	0.0207 ± 0.0001	60.0 ± 0.3
GSU17I	0.4	0.23	0.45	0.75	0.3	80	80	400	20.6	0.0203 ± 0.0003	61.0 ± 0.8
GSU17II	0.4	0.19	0.33	0.75	0.39	80	80	400	28.9	0.0217 ± 0.0001	57.0 ± 0.3
GSU17III	0.4	0.19	0.21	0.75	0.39	80	80	400	37.1	0.0214 ± 0.0001	58.0 ± 0.3

The role of barrier parameters is very critical in determining the photoresponse of extended wavelength detectors. In this view, the effect of δE_V and gradient on the extended threshold wavelength of infrared photodetectors for the temperature range up to 50 K was studied by Somvanshi et al. [103]. In this study, it is observed the δE_V is critical to obtain extended wavelength response λ_{eff} ($\gg \lambda_t$) in IR detectors at 5.3 K; however, the gradient is needed to obtain λ_{eff} at 50 K. A conventional detector shows $\lambda_t - \lambda_{eff}$ over operating temperature from 5.3 K and 50 K whereas the flat injector barrier and barrier-energy-offset-detector (SP1001) shows λ_{eff} ~36 μm at 5.3 K and 4.1 μm at 50 K. When the injector barrier changes from flat (SP1001) to graded (SP1007), the λ_{eff} increases from 4.1 to 8.1 μm at 50 K for a given graded injector barrier, as δE_V increases from 0.10 to 0.19 eV, the λ_{eff} also increases from 8.9 to 13.7 μm. The results of this study clearly indicate that by the optimization of δE_V and the gradient, infrared detectors with λ_{eff} ($\gg \lambda_t$) can be designed to operate over a wide range of temperature.

For practical applications, to provide an advantage over the conventional detector, an extended wavelength detector should have a lower dark current and specific detectivity (D^*) that is comparable or better than the conventional detector. In general, by lowering the operating temperature, the dark current can be reduced; however, this makes detectors' operation more costly. Therefore, lowering the dark current without cooling will be an advantage, especially for longer wavelength detectors for next-generation optoelectronic devices. In extended wavelength detectors, the dark is determined by designed Δ as in the conventional photodetectors. This leads to an advantage in specific detectivity (D^*) for extended wavelength detectors even though their responsivity is much lower as compared to conventional detectors. Using this idea, standard threshold semiconductor detectors could be designed to operate as long-wavelength detectors with a higher value of detectivity and reduced dark current (corresponding to the original short-wavelength threshold).

2. Conclusions

Objects having temperatures higher than absolute zero ($T > 0$ K) radiates the energy in the form of electromagnetic waves and so are the sources of infrared. Thus, infrared detectors can have diverse civilian and military activities applications. Major applications include: estimating heat losses in buildings, roads or any heat-emitting objects of engineering use; imaging and quality control check in biomedical use; security applications for firefighters, night vision, airports; technological applications for electrical circuit manufacturing or identifying faulty connections; application for astronomical studies and much more.

To date, many materials, and physics characteristics have been investigated in IR detection [60,104]. The example includes [105] thermoelectric power (thermocouples) [106], change in electrical conductivity (bolometers) [107], gas expansion (Golay cell) [108], pyroelectricity (pyroelectric detectors) [109], photon drag, Josephson effect (Josephson junctions) [110], internal emission (PtSi Schottky barriers) [111], fundamental absorption (intrinsic photodetectors) [112], impurity absorption (extrinsic photodetectors), low-dimensional solids (superlattice (SL) and quantum well (QW) detectors) [113], and so on. Progress in semiconductor infrared detector technology has been made in IR detector technology to improve their performance. Furthermore, studies are going on for the improvement of detector performances. In reference to this, potential applications of a heterostructure IR detector working within a wide range of temperatures, including a room-temperature environment, can be anticipated. Of course, innovation in the heterostructure architecture of thoroughly documented semiconductor compounds will create another degree of freedom in recent technology.

This study documented the present studies on the S-O band energy, and H-H/L-H transitions give rise to the photoresponse in heterostructure infrared detectors with a peak in the 3–5 μm regime, with the most important band being an atmospheric window. The incorporation of current blocking barriers to lower the dark current density and to increase the performance of heterostructure semiconductor detectors is also discussed. The importance of the wavelength threshold extension mechanism is also clearly illustrated, with its application for future IR photodetector design and development. Finally, a considerable achievement in the practical application of a new kind (unlike the conventional) of IR photodetectors, with the wavelength threshold determined in the near future, can be anticipated.

Photovoltaic infrared sensor arrays for the 3–5 μm and 8–12 μm atmospheric windows can be fabricated using similar techniques. GaAs/AlGaAs heterostructures can overcome the serious technological problem for the mass production of HgCdTe large-sized Focal Plane Arrays (FPA) [114], and to achieve high FPA pixel uniformity and yield. Large line or area arrays of such photovoltaic infrared detectors can be desired for thermal imaging and spectroscopic applications.

Author Contributions: Conceptualization, A.G.U.P.; writing—original draft preparation, H.G.; supervision, A.G.U.P.; writing—review and editing, P.V.V.J., D.S., Y.L., and A.G.U.P. All authors have read and agreed to the published version of the manuscript.

Funding: U.S. Army Research Office (ARO) (W911 NF-15-1-0018).

Acknowledgments: H. Ghimire acknowledges support from Molecular Basis of Disease (MBD) at GSU.

Conflicts of Interest: The authors declare no conflict of interest.

References

1. Weisbuch, C.; Vinter, B. *Quantum Semiconductor Structures: Fundamentals and Applications*; Elsevier: Amsterdam, The Netherlands, 2014.
2. Bhimanapati, G.R.; Lin, Z.; Meunier, V.; Jung, Y.; Cha, J.; Das, S.; Xiao, D.; Son, Y.; Strano, M.S.; Cooper, V.R.; et al. Recent Advances in Two-Dimensional Materials beyond Graphene. *ACS Nano* **2015**, *9*, 11509–11539. [CrossRef]
3. Petroff, P.; DenBaars, S. MBE and MOCVD growth and properties of self-assembling quantum dot arrays in III-V semiconductor structures. *Superlattices Microstruct.* **1994**, *15*, 15. [CrossRef]
4. Neamen, D.A. *Semiconductor Physics and Devices: Basic Principles*; McGraw-Hill: New York, NY, USA, 2012.
5. Nguyen, B.-M.; Hoffman, D.; Delaunay, P.-Y.; Razeghi, M. Dark current suppression in type II InAs/GaSb superlattice long wavelength infrared photodiodes with M-structure barrier. *Appl. Phys. Lett.* **2007**, *91*, 163511. [CrossRef]
6. Krishna, S. Quantum dots-in-a-well infrared photodetectors. *J. Phys. D Appl. Phys.* **2005**, *38*, 2142. [CrossRef]
7. Karunasiri, R.; Park, J.; Wang, K. Si1−x Ge x/Si multiple quantum well infrared detector. *Appl. Phys. Lett.* **1991**, *59*, 2588–2590. [CrossRef]
8. Gunapala, S.D.; Bandara, S.V.; Singh, A.; Liu, J.K.; Rafol, B.; Luong, E.; Mumolo, J.M.; Tran, N.Q.; Ting, D.Z.-Y.; Vincent, J.D.; et al. 640/spl times/486 long-wavelength two-color GaAs/AlGaAs quantum well infrared photodetector (QWIP) focal plane array camera. *IEEE Trans. Electron Devices* **2000**, *47*, 963–971. [CrossRef]
9. Chang, K.H.; Bhattacharya, P.K.; Gibala, R. Characteristics of dislocations at strained heteroepitaxial InGaAs/GaAs interfaces. *J. Appl. Phys.* **1989**, *66*, 2993–2998. [CrossRef]
10. Aziziyan, M.R.; Hassen, W.M. Electrically biased GaAs/AlGaAs heterostructures for enhanced detection of bacteria. *Synth. Photonics Nanoscale Mater. Xiii* **2016**, *9737*, 97370. [CrossRef]
11. Dingle, R.; Wiegmann, W.; Henry, C.H. Quantum states of confined carriers in very thin Al x Ga 1− x As-GaAs-Al x Ga 1− x As heterostructures. *Phys. Rev. Lett.* **1974**, *33*, 827. [CrossRef]

12. Adachi, S. GaAs, AlAs, and Al x Ga1− x As: Material parameters for use in research and device applications. *J. Appl. Phys.* **1985**, *58*, R1–R29. [CrossRef]
13. Yablonovitch, E.; Kapon, E.; Gmitter, T.; Yun, C.; Bhat, R. Double heterostructure GaAs/AlGaAs thin film diode lasers on glass substrates. *IEEE Photon Technol. Lett.* **1989**, *1*, 41–42. [CrossRef]
14. Tomioka, K.; Motohisa, J.; Hara, S.; Hiruma, K.; Fukui, T. GaAs/AlGaAs Core Multishell Nanowire-Based Light-Emitting Diodes on Si. *Nano Lett.* **2010**, *10*, 1639–1644. [CrossRef]
15. Lade, S.J.; Zahedi, A. A revised ideal model for AlGaAs/GaAs quantum well solar cells. *Microelectron. J.* **2004**, *35*, 401–410. [CrossRef]
16. Chauhan, D.; Perera, A.; Li, L.; Chen, L.; Linfield, E. Effect of a current blocking barrier on a 2–6 μ m p-GaAs/AlGaAs heterojunction infrared detector. *Appl. Phys. Lett.* **2016**, *108*, 201105. [CrossRef]
17. Gunapala, S.D.; Bundara, S.; Liu, J.K.; Hong, W.; Sundaram, M.; Maker, P.D.; Muller, R.E.; Shott, C.A.; Carralejo, R. Long-wavelength 640/spl times/486 GaAs-AlGaAs quantum well infrared photodetector snap-shot camera. *IEEE Trans. Electron. Devices* **1998**, *45*, 1890–1895. [CrossRef]
18. Gunapala, S.D.; Bandara, S.V.; Liu, J.K.; Luong, E.; Stetson, N.; Shott, C.A.; Bock, J.J.; Rafol, S.B.; Mumolo, J.M.; McKelvey, M.J. Long-wavelength 256/spl times/256 GaAs/AlGaAs quantum well infrared photodetector (QWIP) palm-size camera. *IEEE Trans. Electron. Devices* **2000**, *47*, 326–332. [CrossRef]
19. Weerasekara, A.; Rinzan, M.; Matsik, S.; Perera, A.G.; Buchanan, M.; Liu, H.C.; Von Winckel, G.; Stintz, A.; Krishna, S. n-Type GaAs/AlGaAs heterostructure detector with a 32 THz threshold frequency. *Opt. Lett.* **2007**, *32*, 1335–1337. [CrossRef]
20. Dai, X.; Zhang, S.; Wang, Z.; Adamo, G.; Liu, H.; Huang, Y.; Couteau, C.; Soci, C. GaAs/AlGaAs Nanowire Photodetector. *Nano Lett.* **2014**, *14*, 2688–2693. [CrossRef]
21. Casey, H.C.; Panish, M.B. *Heterostructure Lasers: Materials and Operating Characteristics*; Academic Pr: Cambridge, MA, USA, 1978.
22. Chang, K.H.; Lee, C.P.; Wu, J.S.; Liu, D.G.; Liou, D.C.; Wang, M.H.; Chen, L.J.; Marais, M.A. Precise determination of aluminum content in AlGaAs. *J. Appl. Phys.* **1991**, *70*, 4877–4882. [CrossRef]
23. Esaev, D.; Rinzan, M.; Matsik, S.; Perera, A. Design and optimization of GaAs/AlGaAs heterojunction infrared detectors. *J. Appl. Phys.* **2004**, *96*, 4588–4597. [CrossRef]
24. Rogalski, A.; Chrzanowski, K. 18 Infrared devices and techniques. In *Nanophotonics and Plasmonics*; Informa UK Ltd.: Colchester, UK, 2017; pp. 633–686.
25. Rogalski, A. *Infrared and Terahertz Detectors*; CRC Press: Boca Raton, FL, USA, 2019.
26. Coon, D.D.; Karunasiri, R.P.G. New mode of IR detection using quantum wells. *Appl. Phys. Lett.* **1984**, *45*, 649. [CrossRef]
27. Rogalski, A. Heterostructure infrared photovoltaic detectors. *Infrared Phys. Technol.* **2000**, *41*, 213–238. [CrossRef]
28. Ariyawansa, G.; Rinzan, M.B.M.; Alevli, M.; Strassburg, M.; Dietz, N.; Perera, A.G.U.; Matsik, S.G.; Asghar, A.; Ferguson, I.T.; Luo, H.; et al. GaN/AlGaN ultraviolet/infrared dual-band detector. *Appl. Phys. Lett.* **2006**, *89*, 091113. [CrossRef]
29. Perera, A.G.U. Heterojunction and superlattice detectors for infrared to ultraviolet. *Prog. Quantum Electron.* **2016**, *48*, 1–56. [CrossRef]
30. Tong, J.; Tobing, L.Y.; Qiu, S.; Zhang, D.H.; Perera, A.U. Room temperature Plasmon-enhanced InAs0. 91Sb0. 09-based heterojunction nip mid-wave infrared photodetector. *Appl. Phys. Lett.* **2018**, *113*, 011110. [CrossRef]
31. Kouvetakis, J.; Chizmeshya, A.V.G. New classes of Si-based photonic materials and device architectures via designer molecular routes. *J. Mater. Chem.* **2007**, *17*, 1649. [CrossRef]
32. Lin, Y.; Donetsky, D.; Wang, D.; Westerfeld, D.; Kipshidze, G.; Shterengas, L.; Sarney, W.L.; Svensson, S.P.; Belenky, G. Development of Bulk InAsSb Alloys and Barrier Heterostructures for Long-Wave Infrared Detectors. *J. Electron. Mater.* **2015**, *44*, 3360–3366. [CrossRef]
33. Tsaur, B.-Y.; Chen, C.; Marino, S. Long-wavelength Ge/sub x/Si/sub 1-x//Si heterojunction infrared detectors and 400*400-element imager arrays. *IEEE Electron. Device Lett.* **1991**, *12*, 293–296. [CrossRef]
34. Harris, D.C. Durable 3–5 μm transmitting infrared window materials. *Infrared Phys. Technol.* **1998**, *39*, 185–201. [CrossRef]
35. Salisbury, J.W.; D'Aria, D.M. Emissivity of terrestrial materials in the 8–14 μm atmospheric window. *Remote Sens. Environ.* **1992**, *42*, 83–106. [CrossRef]

36. Bastard, G.; Schulman, J. Wave Mechanics Applied to Semiconductor Heterostructures. *Phys. Today* **1992**, *45*, 103. [CrossRef]
37. Moreau, W.M. *Semiconductor Lithography: Principles, Practices, and Materials*; Springer Science & Business Media: Berlin, Germany, 2012.
38. Pelzel, R. A comparison of MOVPE and MBE growth technologies for III-V epitaxial structures. In Proceedings of the CS MANTECH Conference, New Orleans, LA, USA, 13–16 May 2013; pp. 105–108.
39. Tobin, S.; Vernon, S.; Bajgar, C.; Wojtczuk, S.; Melloch, M.; Keshavarzi, A.; Stellwag, T.; Venkatensan, S.; Lundström, M.; Emery, K. Assessment of MOCVD- and MBE-growth GaAs for high-efficiency solar cell applications. *IEEE Trans. Electron. Devices* **1990**, *37*, 469–477. [CrossRef]
40. Horikoshi, Y.; Kawashima, M.; Yamaguchi, H. Migration-Enhanced Epitaxy of GaAs and AlGaAs. *Jpn. J. Appl. Phys.* **1988**, *27*, 169–179. [CrossRef]
41. Lundstrom, M.; Schuelke, R. Numerical analysis of heterostructure semiconductor devices. *IEEE Trans. Electron. Devices* **1983**, *30*, 1151–1159. [CrossRef]
42. Tersoff, J. Theory of semiconductor heterojunctions: The role of quantum dipoles. *Phys. Rev. B* **1984**, *30*, 4874–4877. [CrossRef]
43. Capasso, F.; Cho, A.Y. Bandgap engineering of semiconductor heterostructures by molecular beam epitaxy: Physics and applications. *Surf. Sci.* **1994**, *299*, 878–891. [CrossRef]
44. Liu, Q.; Zhang, X.; Abdalla, L.B.; Zunger, A. Transforming Common III-V and II-VI Semiconductor Compounds into Topological Heterostructures: The Case of CdTe/InSb Superlattices. *Adv. Funct. Mater.* **2016**, *26*, 3259–3267. [CrossRef]
45. Chang, K.H.; Gilbala, R.; Srolovitz, D.J.; Bhattacharya, P.K.; Mansfield, J.F. Crosshatched surface morphology in strained III-V semiconductor films. *J. Appl. Phys.* **1990**, *67*, 4093–4098. [CrossRef]
46. Ting, D.Z.; Soibel, A.; Hill, C.; Nguyen, J.; Keo, S.; Rafol, S.B.; Yang, B.; Lee, M.C.; Mumolo, J.; Liu, J.; et al. Antimonide superlattice complementary barrier infrared detector (CBIRD). *Infrared Phys. Technol.* **2011**, *54*, 267–272. [CrossRef]
47. Lin, T.; Maserjian, J. Novel Si1− x Ge x/Si heterojunction internal photoemission long-wavelength infrared detectors. *Appl. Phys. Lett.* **1990**, *57*, 1422–1424. [CrossRef]
48. Yu, J.S.; Ko, Y.H.; Nagaraju, G. Low-dimensional II-VI oxide-based semiconductor nanostructure photodetectors for light sensing. *Quantum Sens. Nanophotonic Devices Xii* **2015**, *9370*, 93702. [CrossRef]
49. Piotrowski, J.; Rogalski, A. New generation of infrared photodetectors. *Sens. Actuators A Phys.* **1998**, *67*, 146–152. [CrossRef]
50. Rogalski, A. Recent progress in infrared detector technologies. *Infrared Phys. Technol.* **2011**, *54*, 136–154. [CrossRef]
51. Rogalski, A.; Adamiec, K.; Rutkowski, J. *Narrow-Gap Semiconductor Photodiodes*; SPIE Press: Bellingham, WA, USA, 2000; Volume 77.
52. Bratt, P.R. Potential barriers in HgCdTe heterojunctions. *J. Vac. Sci. Technol. A* **1985**, *3*, 238–245. [CrossRef]
53. Rogalski, A. Infrared detectors: Status and trends. *Prog. Quantum Electron.* **2003**, *27*, 59–210. [CrossRef]
54. Levine, B.F. Quantum-well infrared photodetectors. *J. Appl. Phys.* **1993**, *74*, 1. [CrossRef]
55. Krishna, S.; Gunapala, S.; Bandara, S.; Hill, C.; Ting, D. Quantum Dot Based Infrared Focal Plane Arrays. *Proc. IEEE* **2007**, *95*, 1838–1852. [CrossRef]
56. Jones, R.C. Performance of Detectors for Visible and Infrared Radiation. In *Advances in Electronics and Electron Physics*; Elsevier B.V.: Danvers, MA, USA, 1953; Volume 5, pp. 1–96.
57. Paschotta, R. Avalanche photodiodes. *Encycl. Laser Phys. Technol.* 2008. Available online: https://www.rp-photonics.com/avalanche_photodiodes.html (accessed on 21 April 2020).
58. Field, H. UV-VIS-IR spectral responsivity measurements system for solar cells. In Proceedings of the National Center for photovoltaics (NCPV) 15th Program Review Meeting, Denver, CO, USA, 9–11 September 1998; pp. 629–635. [CrossRef]
59. Andersson, J.Y.; Lundqvist, L. Near?unity quantum efficiency of AlGaAs/GaAs quantum well infrared detectors using a waveguide with a doubly periodic grating coupler. *Appl. Phys. Lett.* **1991**, *59*, 857–859. [CrossRef]
60. Dereniak, E.L.; Boreman, G.D. *Infrared Detectors and Systems*; Wiley: New York, NY, USA, 1996; Volume 306.

61. Cervera, C.; Baier, N.; Gravrand, O.; Mollard, L.; Lobre, C.; Destéfanis, G.; Zanatta, J.P.; Boulade, O.; Moreau, V. Low-dark current p-on-n MCT detector in long and very long-wavelength infrared. *Infrared Technol. Appl. Xli* **2015**, *9451*, 945129. [CrossRef]
62. Chauhan, D.; Perera, A.G.U.; Li, L.; Chen, L.; Linfield, E.H. Dark current and photoresponse characteristics of extended wavelength infrared photodetectors. *J. Appl. Phys.* **2017**, *122*, 024501. [CrossRef]
63. Weber, E.R.; Willardson, R.K.; Liu, H.; Capasso, F. *Intersubband Transitions in Quantum Wells: Physics and Device Applications*; Academic press: San Diego, CA, USA, 1999.
64. Bie, Y.-Q.; Liao, Z.-M.; Zhang, H.; Li, G.-R.; Ye, Y.; Zhou, Y.; Xu, J.; Qin, Z.-X.; Dai, L.; Yu, D. Self-Powered, Ultrafast, Visible-Blind UV Detection and Optical Logical Operation based on ZnO/GaN Nanoscale p-n Junctions. *Adv. Mater.* **2010**, *23*, 649–653. [CrossRef] [PubMed]
65. Vasko, F.T.; Kuznetsov, A.V. *Electronic States and Optical Transitions in Semiconductor Heterostructures*; Springer Science and Business Media LLC: Berlin, Germany, 1999.
66. Souri, D.; Tahan, Z.E. A new method for the determination of optical band gap and the nature of optical transitions in semiconductors. *Appl. Phys. A* **2015**, *119*, 273–279. [CrossRef]
67. Zhuravlev, K.S.; A Kolosanov, V.; Milekhin, A.G.; Polovinkin, V.; Shamirzaev, T.S.; Rakov, Y.N.; Myakishev, Y.B.; Fryar, J.; McGlynn, E.; O Henry, M. Infrared light emission from GaAs MESFETs operating at avalanche breakdown conditions. *Semicond. Sci. Technol.* **2004**, *19*, S94–S95. [CrossRef]
68. Hoff, J.; Razeghi, M.; Brown, G.J. Effect of the spin split-off band on optical absorption in p-type Ga 1− x In x As y P 1− y quantum-well infrared detectors. *Phys. Rev. B* **1996**, *54*, 10773. [CrossRef] [PubMed]
69. Lao, Y.F.; Perera, A.G.U.; Wang, H.L.; Zhao, J.; Jin, Y.; Zhang, D.H. Optical characteristics of p-type GaAs-based semiconductors towards applications in photoemission infrared detectors. *J. Appl. Phys.* **2016**, *119*, 105304. [CrossRef]
70. Perera, A.G.U.; Matsik, S.; Jayaweera, P.V.V.; Tennakone, K.; Liu, H.C.; Buchanan, M.; Von Winckel, G.; Stintz, A.; Krishna, S. High operating temperature split-off band infrared detectors. *Appl. Phys. Lett.* **2006**, *89*, 131118. [CrossRef]
71. Jayaweera, P.; Matsik, S.; Tennakone, K.; Perera, A.G.U.; Liu, H.; Krishna, S. Spin split-off transition based IR detectors operating at high temperatures. *Infrared Phys. Technol.* **2007**, *50*, 279–283. [CrossRef]
72. Perera, A.G.U.; Jayaweera, P.; Matsik, S.; Liu, H.; Buchanan, M.; Wasilewski, Z. Operating temperature and the responsivity of split-off band detectors. *Infrared Phys. Technol.* **2009**, *52*, 241–246. [CrossRef]
73. Esaev, D.G.; Rinzan, M.B.M.; Matsik, S.; Perera, A.G.U.; Liu, H.C.; Zvonkov, B.N.; Gavrilenko, V.I.; Belyanin, A.A. High performance single emitter homojunction interfacial work function far infrared detectors. *J. Appl. Phys.* **2004**, *95*, 512–519. [CrossRef]
74. Perera, A.G.U.; Shen, W.; Liu, H.; Buchanan, M.; Schaff, W. GaAs homojunction interfacial workfunction internal photoemission (HIWIP) far-infrared detectors. *Mater. Sci. Eng. B* **2000**, *74*, 56–60. [CrossRef]
75. Jayaweera, P.V.V.; Matsik, S.G.; Perera, A.G.U.; Liu, H.C.; Buchanan, M.; Wasilewski, Z.R. Uncooled infrared detectors for 3–5μm and beyond. *Appl. Phys. Lette.* **2008**, *93*, 021105. [CrossRef]
76. Lao, Y.; Pitigala, P.K.D.D.; Perera, A.G.U.; Liu, H.C.; Buchanan, M.; Wasilewski, Z.R.; Choi, K.; Wijewarnasuriya, P. Light-hole and heavy-hole transitions for high-temperature long-wavelength infrared detection. *Appl. Phys. Lett.* **2010**, *97*, 091104. [CrossRef]
77. Matsik, S.; Jayaweera, P.V.V.; Perera, A.G.U.; Choi, K.; Wijewarnasuriya, P. Device modeling for split-off band detectors. *J. Appl. Phys.* **2009**, *106*, 064503. [CrossRef]
78. Perera, A.G.U.; Jayaweera, P.; Ariyawansa, G.; Matsik, S.; Tennakone, K.; Buchanan, M.; Liu, H.; Su, X.; Bhattacharya, P. Room temperature nano- and microstructure photon detectors. *Microelectron. J.* **2009**, *40*, 507–511. [CrossRef]
79. Pitigala, P.K.D.D.; Matsik, S.; Perera, A.G.U.; Khanna, S.; Li, L.; Linfield, E.H.; Wasilewski, Z.R.; Buchanan, M.; Liu, H.C. Photovoltaic infrared detection with p-type graded barrier heterostructures. *J. Appl. Phys.* **2012**, *111*, 084505. [CrossRef]
80. Perera, A.G.U.; Matsik, S.G.; Yaldiz, B.; Liu, H.C.; Shen, A.; Gao, M.; Wasilewski, Z.R.; Buchanan, M. Heterojunction wavelength-tailorable far-infrared photodetectors with response out to 70 μm. *Appl. Phys. Lett.* **2001**, *78*, 2241–2243. [CrossRef]
81. Wang, S.Y.; Lin, S.D.; Wu, H.W.; Lee, C.P. Low dark current quantum-dot infrared photodetectors with an AlGaAs current blocking layer. *Appl. Phys. Lett.* **2001**, *78*, 1023–1025. [CrossRef]

82. Rotella, P.; Raghavan, S.; Stintz, A.; Fuchs, B.; Krishna, S.; Morath, C.; Le, D.; Kennerly, S. Normal incidence InAs/InGaAs dots-in-well detectors with current blocking AlGaAs layer. *J. Cryst. Growth* **2003**, *251*, 787–793. [CrossRef]
83. Wang, S.; Lin, S.; Wu, H.; Lee, C. High performance InAs/GaAs quantum dot infrared photodetectors with AlGaAs current blocking layer. *Infrared Phys. Technol.* **2001**, *42*, 473–477. [CrossRef]
84. Lin, S.-Y.; Tsai, Y.-R.; Lee, S.-C. High-performance InAs/GaAs quantum-dot infrared photodetectors with a single-sided Al0.3Ga0.7As blocking layer. *Appl. Phys. Lett.* **2001**, *78*, 2784–2786. [CrossRef]
85. Pal, D.; Towe, E. Uniformly doped InAs/GaAs quantum-dot infrared photodetectors with AlGaAs current blocking layer. In Proceedings of the 5th IEEE Conference on Nanotechnology, Nagoya, Japan, 15 July 2005; Institute of Electrical and Electronics Engineers (IEEE): Piscataway, NJ, USA, 2005; pp. 418–421.
86. Nevou, L.; Liverini, V.; Castellano, F.; Bismuto, A.; Faist, J. Asymmetric heterostructure for photovoltaic InAs quantum dot infrared photodetector. *Appl. Phys. Lett.* **2010**, *97*, 23505. [CrossRef]
87. Stiff, A.; Krishna, S.; Bhattacharya, P.; Kennerly, S. Normal-incidence, high-temperature, mid-infrared, InAs-GaAs vertical quantum-dot infrared photodetector. *IEEE J. Quantum Electron.* **2001**, *37*, 1412–1419. [CrossRef]
88. Tang, S.-F.; Lin, S.-Y.; Lee, S.-C. Near-room-temperature operation of an InAs/GaAs quantum-dot infrared photodetector. *Appl. Phys. Lett.* **2001**, *78*, 2428–2430. [CrossRef]
89. Chakrabarti, S.; Stiff-Roberts, A.; Bhattacharya, P.; Gunapala, S.; Bandara, S.; Rafol, S.; Kennerly, S. High-Temperature Operation of InAs–GaAs Quantum-Dot Infrared Photodetectors With Large Responsivity and Detectivity. *IEEE Photon Technol. Lett.* **2004**, *16*, 1361–1363. [CrossRef]
90. Nguyen, B.-M.; Bogdanov, S.; Pour, S.A.; Razeghi, M. Minority electron unipolar photodetectors based on type II InAs/GaSb/AlSb superlattices for very long wavelength infrared detection. *Appl. Phys. Lett.* **2009**, *95*, 183502. [CrossRef]
91. Ting, D.Z.; Hill, C.; Soibel, A.; Keo, S.; Mumolo, J.; Nguyen, J.; Gunapala, S.D. A high-performance long wavelength superlattice complementary barrier infrared detector. *Appl. Phys. Lett.* **2009**, *95*, 023508. [CrossRef]
92. Gautam, N.; Kim, H.S.; Kutty, M.N.; Plis, E.; Dawson, L.R.; Krishna, S. Performance improvement of longwave infrared photodetector based on type-II InAs/GaSb superlattices using unipolar current blocking layers. *Appl. Phys. Lett.* **2010**, *96*, 231107. [CrossRef]
93. Somvanshi, D.; Chauhan, D.; Lao, Y.-F.; Perera, A.G.U.; Li, L.; Khanna, S.; Linfield, E.H. Analysis of Extended Threshold Wavelength Photoresponse in Nonsymmetrical p-GaAs/AlGaAs Heterostructure Photodetectors. *IEEE J. Sel. Top. Quantum Electron.* **2018**, *24*, 1–7. [CrossRef]
94. Shah, J. Hot electrons and phonons under high intensity photoexcitation of semiconductors. *Solid-State Electron.* **1978**, *21*, 43–50. [CrossRef]
95. Ulbrich, R. Low density photoexcitation phenomena in semiconductors: Aspects of theory and experiment. *Solid-State Electron.* **1978**, *21*, 51–59. [CrossRef]
96. Shah, J. Investigation of hot carrier relaxation with picosecond laser pulses. *Le J. De Phys. Colloq.* **1981**, *42*, 7. [CrossRef]
97. Hayes, J.R.; Levi, A.F.J.; Wiegmann, W. Hot-Electron Spectroscopy of GaAs. *Phys. Rev. Lett.* **1985**, *54*, 1570–1572. [CrossRef] [PubMed]
98. Shah, J. Hot carriers in quasi-2-D polar semiconductors. *IEEE J. Quantum Electron.* **1986**, *22*, 1728–1743. [CrossRef]
99. Chauhan, D.; Perera, A.U.; Li, L.; Chen, L.; Khanna, S.; Linfield, E.H. Extended wavelength infrared photodetectors. *Opt. Eng.* **2017**, *56*, 091605. [CrossRef]
100. Chauhan, D.; Perera, A.G.U.; Li, L.; Chen, L.; Linfield, E. Study of infrared photodetectors with wavelength extension mechanism. *Infrared Phys. Technol.* **2018**, *95*, 148–151. [CrossRef]
101. Chauhan, D.; Perera, A.G.U.; Li, L.; Chen, L.; Linfield, E.H. Effects of Barrier Energy Offset and Gradient in Extended Wavelength Infrared Detectors. *IEEE Sens. Lett.* **2018**, *2*, 1–4. [CrossRef]
102. Somvanshi, D.; Chauhan, D.; Perera, A.G.U.; Li, L.; Chen, L.; Linfield, E.H. Reduced Dark Current With a Specific Detectivity Advantage in Extended Threshold Wavelength Infrared Detector. *IEEE Sens. Lett.* **2019**, *3*, 1–4. [CrossRef]

103. Somvanshi, D.; Chauhan, D.; Perera, A.G.U.; Li, L.; Chen, L.; Linfield, E.H. Analysis of Barrier Parameters on the Extended Threshold Wavelength of Infrared Detectors. *IEEE Photon Technol. Lett.* **2018**, *30*, 1617–1620. [CrossRef]
104. Norton, P. Infrared detectors in the next millennium. *Aerosense '99* **1999**, *3698*, 652–665. [CrossRef]
105. Rogalski, A. Infrared detectors: An overview. *Infrared Phys. Technol.* **2002**, *43*, 187–210. [CrossRef]
106. Szakmany, G.P.; Krenz, P.M.; Orlov, A.O.; Bernstein, G.H.; Porod, W. Antenna-coupled nanowire thermocouples for infrared detection. *IEEE Trans. Nanotechnol.* **2012**, *12*, 163–167. [CrossRef]
107. Rogalski, A. *Infrared Detectors*; CRC Press: Boca Raton, FL, USA, 2010.
108. Fernandes, L.O.T.; Kaufmann, P.; Marcon, R.; Kudaka, A.S.; Marun, A.; Godoy, R.; Bortolucci, E.C.; Zakia, M.B.; Diniz, J.A. Photometry of THz radiation using Golay cell detector. *2011 XXXth URSI Gen. Assem. Sci. Symp.* **2011**, 1–4. [CrossRef]
109. Shankar, M.; Burchett, J.; Hao, Q.; Guenther, B.D.; Brady, D.J. Human-tracking systems using pyroelectric infrared detectors. *Opt. Eng.* **2006**, *45*, 106401. [CrossRef]
110. Gallop, J.C. *SQUIDs, the Josephson Effects and Superconducting Electronics*; Informa UK Ltd.: Colchester, UK, 2017.
111. Raissi, F.; Far, M. Highly sensitive PtSi/porous Si Schottky detectors. *IEEE Sens. J.* **2002**, *2*, 476–481. [CrossRef]
112. Urich, A.; Unterrainer, K.; Mueller, T. Intrinsic Response Time of Graphene Photodetectors. *Nano Lett.* **2011**, *11*, 2804–2808. [CrossRef] [PubMed]
113. Ellis, A.R.; Reno, J.L.; Tsui, D.C.; Majumdar, A.; Choi, K. Binary superlattice quantum-well infrared photodetectors for long-wavelength broadband detection. *Appl. Phys. Lett.* **2004**, *84*, 5127. [CrossRef]
114. Lewis, E.N.; Treado, P.J.; Reeder, R.C.; Story, G.M.; Dowrey, A.E.; Marcott, C.; Levin, I.W. Fourier Transform Spectroscopic Imaging Using an Infrared Focal-Plane Array Detector. *Anal. Chem.* **1995**, *67*, 3377–3381. [CrossRef]

© 2020 by the authors. Licensee MDPI, Basel, Switzerland. This article is an open access article distributed under the terms and conditions of the Creative Commons Attribution (CC BY) license (http://creativecommons.org/licenses/by/4.0/).

MDPI
St. Alban-Anlage 66
4052 Basel
Switzerland
Tel. +41 61 683 77 34
Fax +41 61 302 89 18
www.mdpi.com

Micromachines Editorial Office
E-mail: micromachines@mdpi.com
www.mdpi.com/journal/micromachines

CPSIA information can be obtained
at www.ICGtesting.com
Printed in the USA
LVHW070558230322
714166LV00009B/330

9 783036 533537